Brexit and the Consequences for International Competitiveness

Arkadiusz Michał Kowalski
Editor

Brexit and the Consequences for International Competitiveness

palgrave
macmillan

Editor
Arkadiusz Michał Kowalski
Collegium of World Economy
SGH Warsaw School of Economics
Warsaw, Poland

ISBN 978-3-030-03244-9 ISBN 978-3-030-03245-6 (eBook)
https://doi.org/10.1007/978-3-030-03245-6

Library of Congress Control Number: 2018960884

Cover illustration: Boschettophotography/Getty Images

This Palgrave Macmillan imprint is published by the registered company Springer Nature Switzerland AG
The registered company address is: Gewerbestrasse 11, 6330 Cham, Switzerland

This monograph is the result of statutory research that has been financed by Polish Ministry of Science and Higher Education and conducted by the research team of the World Economic Research Institute of the Collegium of World Economy at the Warsaw School of Economics.

Preface

Many theoretical and empirical studies have been conducted in recent decades in response to deepening economic integration processes in Europe and elsewhere. Meanwhile, the 23 June 2016 referendum in the UK showed that regional integration does not have to be a one-way process as it was once thought to be. The key objective of this publication is to identify the implications of Brexit for structural changes in the global economy, taking into account selected aspects of the disintegration process which started in Europe in 2016.

This book starts with a chapter by Arkadiusz Michał Kowalski that provides a theoretical background to the Brexit process and aims to contribute to developing the theory of regional economic disintegration. The chapter analyses the consequences of Britain's impending departure from the European Union for the international competitiveness of economies, including with respect to international trade and foreign direct investment (FDI). The traditional view of integration as a process where countries deepen cooperation and subsequently switch to modes involving stronger commitment—starting from a free-trade area, followed by a custom union, a common market, an economic union and, finally, complete integration—is presented. However, the author shows that this process may be reversed and turned into regional economic disintegration. Moreover, theoretical insights are discussed with respect to regional disintegration coming from the concepts of new intergovernmentalism and neo-functionalism.

The main objective of the next chapter, "Brexit and Innovation: Focus on Research and Development in the UK," by Marzenna Anna Weresa, is to identify how the UK's position will change after Brexit in terms of participation in EU-funded research and innovation programmes. The UK has a strong science base, and the country's innovation performance, as measured by the European Union's Summary Innovation Index, has improved by 11.7% since 2010. The British innovation system is highly internationalised compared to other EU member states. The involvement of UK researchers in European research projects financed through framework programme schemes has been increasing, and the country receives more from the EU's R&D budget than it contributes. The chapter shows that the impact of Brexit on research and innovation in the UK depends on the model of the UK's future relationship with the EU. The most probable scenario for UK–EU research collaboration post-Brexit is that the UK will become a partially associated country. Such a status will mean that the UK will contribute to EU framework programmes based on GDP and population, but will have no role in developing EU research and innovation policy and no influence on funding topics. In the long run, the UK may follow the same scenario as Switzerland and experience a decline in participation and financial benefits received from EU research programmes. This may have a negative impact on the UK's international research collaboration in the post-Brexit period and become a factor limiting further internationalisation of the country's national innovation system.

Chapter 3, entitled "Brexit as a National Transformation Programme: Project Management Perspective," by Ewa Sońta-Drączkowska, discusses Brexit from a project management perspective. The purpose of the chapter is to analyse Brexit as a national transformation programme from the perspective of project management discipline. The key themes of the transformation are analysed as well as the role of project management in this process. Several questions and issues are raised based on programme management methodologies as well as research reports on managing complex projects in the public sector. Special attention is given to two areas: governance issues and civil service resources. The "Managing Successful Programmes" guidance, which is a compendium of programme management best practices, is the analytical framework for the study. Public value management theory (PVM) provides a theoretical perspective for the analysis. The study finds that, from the project management perspective, Brexit can be perceived as a high-risk programme with several shortcomings concerning proper programme set-up and management. The risk increases due to limited capabilities of the British civil service. Therefore, Brexit will probably result in long-lasting projects conducted in the UK public sector for many years to come.

In Chapter 4, "The Economic Order of Post-Brexit Europe and the Role of Germany: An Ordoliberal Perspective," Jürgen Wandel analyses the scenarios of the EU's post-Brexit economic order and the role of Germany in shaping it from an ordoliberal perspective. The analysis shows that the ordoliberal view would necessitate a flexible Europe of different clubs based on a free and open single market rather than an ever closer union. Yet, the author argues that Germany seems unlikely to push for such a solution for two reasons. First, Germany's own economic policy is becoming less oriented towards free markets. Second, the "sacralisation" of the European project in Germany makes it almost taboo to question federalisation as the final goal of European integration.

Chapter 5, "Changes in Germany's European Policy in the Face of Brexit," by Józef Olszyński, highlights the fact that the German leadership of the European Union, which has emerged and been consolidated over the decades, has been confronted with the institutional disintegration caused by Britain's decision to exit the EU in a move known as Brexit. This new phenomenon of decomposition of the European integration process brings—besides economic and non-economic effects for Britain and the whole Union—an additional challenge for Germany. The German economy will be exposed to tangible losses, and Germany's European policy will require difficult adjustments. The improvement of Germany's position in the EU is accompanied by an unequal increase in disparities with France and increased responsibility to meet the expectations of EU partners. Domestic as well as intra-EU and global factors will require the German government to increase the flexibility of "management" and intensify cooperation with EU member states.

Chapter 6, "Bilateral Trade and Investment Between Britain and Germany Ahead of the UK's Impending Departure from the European Union," by Andreas Bielig, shows that the UK's political decision to leave the European Union has influenced the German economy in two main ways. First, uncertainty over the potential future impact of Brexit has forced businesses to adapt to the new situation according to their risk profile. Since the so-called Brexit referendum adaptive arrangements by firms have been increasingly in evidence. Second, even though the actual impact of the exit will depend on its detailed terms and conditions, enterprises need to make early adjustments to maximise their profits, earnings and utility. As long as the exact conditions of Brexit are unclear, only a rough picture can be drawn of the expected adaptations at both the micro- and macroeconomic levels. The author outlines the current state of bilateral trade between Britain and Germany and highlights some potential impacts of Brexit. The chapter uses data from Germany's Destatis Statistical Office on foreign trade and direct investment at the two-digit sector level. In 2016, the UK ranked third among Germany's largest export markets and was No. 11 among the country's top import partners. This means that restrictions in bilateral trade could be particularly painful for German exporters. The author also analyses bilateral investment patterns, which are closely connected with trade, evaluating factors determining potential future changes in UK direct investment in Germany and German investment in Britain.

The aim of Chapter 7, "Economic Consequences of Brexit for Poland," by Mariusz-Jan Radło is to analyse the economic impact of Brexit on the Polish economy. In our analysis, we show that Brexit may have various consequences for the Polish economy depending on the exact shape of post-Brexit economic relations between the EU and the UK. While most macroeconomic analyses suggest that Britain's departure from the EU will have a negative effect on overall GDP growth, the actual impact of Brexit on the Polish economy may be negligible. When it comes to migration, Brexit can have a positive impact on the Polish economy. Nevertheless, a key goal from the point of view of Poland's economic interests should be to use Brexit to reverse migration trends and attract back some of the tens of thousands of Poles living and working in Britain. Thus, Brexit may positively influence Poland's labour market and long-term economic growth. Meanwhile, Brexit may negatively affect trade and foreign investment, though the actual impact may vary depending on the Brexit scenario. Interestingly, Brexit should improve the position of Poland and its regional Visegrad Group partners in the Council of the European Union and the European Council. It will also change the bargaining power of different coalitions of countries, including the possibility of voting or blocking decisions.

Chapter 8, "Economic Implications of Brexit for the International Competitiveness of Russia," by Krzysztof Falkowski, examines the possible impact of the UK's impending EU exit on the international competitiveness of the Russian economy. To this end, an in-depth analysis of Russia's competitive profile is conducted to identify the main advantages and disadvantages of its economy in international trade, followed by a discussion of relations between Russia and the UK in trade, investment and the movement of labour. The analysis shows that, due to the scale of mutual economic cooperation and the existing competitive profile of the Russian economy, Brexit will most likely have a limited impact on Russia's competitiveness in the global economy.

The main aim of Chapter 9, "Trade Openness and FDI in the UK After Brexit," by Tomasz M. Napiórkowski, is to examine the possible effects that the UK's decision to leave the European Union may have on the British economy due to a change in the country's attractiveness to foreign direct investment (FDI). The study focuses on openness to trade as a channel through which Brexit will impact inward FDI activity in the UK. After establishing the benefits of being an FDI host and the role of trade openness as one of the key determinants of inward FDI, the study finds that the relative attractiveness of the UK as a host of FDI has decreased as less of the world's inward FDI is being directed at the UK. With a set of econometric tests, the study shows that the relationship between trade openness and inward FDI for the UK is positive and strong, and that the causality runs from trade openness to inward FDI, but there is no feedback. Therefore, the UK leaving the EU, through a fall in trade openness and therefore a fall in inward FDI, would have a significant negative impact on the UK's economy.

Chapter 10, "Brexit and Britain's Relations with South Korea," by Piotr Ostaszewski, analyses Brexit and its possible consequences for UK–South Korea relations. The first part of the chapter focuses on broader aspects such as economic relations between the UK and the Asia-Pacific area as well as the benefits that some powerful Asian economies might derive from Brexit. The second part traces EU–South Korean relations with Britain as a member country. The third part explores British–South Korean relations prior to the Brexit vote, while part four examines the possibility of new free-trade agreement (FTA) negotiations between the two countries.

Chapter 11, "Does Brexit Influence China's "One Belt One Road" Initiative?" by Günter Heiduk, analyses the Brexit process from a Chinese policy perspective. China's President Xi Jinping's October 2015 visit to the UK was heralded as the start of a "golden age" in UK-China trade and investment relations. Shortly before that, in March 2015, the Chinese government officially launched the country's Belt and Road Initiative (BRI), which aims to mobilise new decade-long growth forces at home and abroad through building a China–Europe rail network and establishing related infrastructure facilities. The 23 June 2016 referendum in the UK resulted in most British citizens voting in favour of an exit from the EU. This raises the question of whether China views Brexit in terms of the "sudden death" of the UK–China "golden age" or whether it believes in a long-term recovery from this shock. Some experts argue that Brexit is a major setback for UK–China relations, while others expect a pragmatic reaction of China's leaders motivated by a long-term geostrategic interest in the UK. With regard to the BRI, Brexit may cause the Chinese government to review the current projects in the UK and, if necessary, adjust them to the expected post-Brexit situation. This might have implications on China's BRI strategy in continental Europe, especially in Central and Eastern European Countries (CEECs). The main aim of the chapter is to put forward plausible arguments on how China might react to Brexit in various key BRI projects. The author concludes that, first, China may push forward with the Arctic Sea route to London, which bypasses the EU27, to at least partly replace the Yiwu–London rail route. Second, with respect to the establishment of new Chinese nuclear power plants in the UK, Britain leaving the European Atomic Energy Community (EURATOM) might create unexpected obstacles in terms of safety and security concerns. Third, a Chinese-built and -owned new business district on the grounds of the London Royal Albert Dock could be seen in Britain as a new kind of dependence triggering Sinophobic sentiment. Fourth, Brexit may hamper the role of the City of London as the leading global financial centre, but this is unlikely to cause a drastic re-orientation in terms of China switching away from London—in favour of another location in Europe—as the key location for financing BRI projects and for internationalising its renminbi currency. There are a number of uncertainties, especially regarding the final post-Brexit arrangements between the major stakeholders, which creates incentives to closely observe developments in post-Brexit China–UK relations, especially with respect to BRI projects.

Chapter 12, "The Impact of Brexit on Foreign Direct Investment and Trade Relations Between the UK and China," by Marta Mackiewiczand Agnieszka McCaleb, highlights the fact that the UK has been attracting the largest chunk of China's outward FDI among EU members. The country has been open to China's economic goals, including a free-trade agreement, market economy status and an investment agreement. With the UK out of the EU, China may face more protectionist attitudes that dominate in continental Europe. With the details of Brexit still being worked out, the chapter investigates the possible implications of Brexit for the international competitiveness of China. It finds that Brexit will have little impact on Chinese FDI flows in the UK as these are mostly motivated by strategic-asset seeking. Due to the British pound losing its value after the Brexit vote, these assets have become cheaper for Chinese investors. A quantitative analysis of trade between the UK and China indicates that, although there is potential for the development and transformation of UK–China trade, it is probably not as large as is often claimed and depends on post-Brexit relations between the EU, the UK and China.

Chapter 13, "The Potential Impacts of Brexit on the Japanese Economy," by Anna Maria Dzienis, aims to shed light on the economic and political consequences of Brexit for EU partner countries. Japan serves as a good example because, according to the UK Office of National Statistics, it was among the top 10 foreign investors in the UK in 2016 in terms of both FDI flows and overall position. The Donald Trump administration's January 2017 decision to withdraw the USA from *Trans-Pacific Partnership* (TPP) negotiations raised fears of protectionism. In this context, the EU–Japan Economic Partnership Agreement, representing nearly 20% of the global GDP, is considered to be the most significant free-trade arrangement in the world today.

Finally, Chapter 14, "Brexit and Sterling Depreciation: Impact on Selected Economies," by Anna Sznajderska, focuses on the depreciation of the British pound sterling as one of the biggest economic consequences of the Brexit decision. The biggest one-day sterling fall against the dollar was observed on the night of the decision itself. The aim of the chapter is to assess the impact of an unexpected depreciation of the pound sterling on economic growth and stock price indices in the UK and selected other economies. Specifically, the author applies a global vector autoregressive model to analyse the whole global economy consisting of many interlinked countries. The study shows that an unexpected drop in the value of the sterling has the potential to reduce economic activity in the UK and a number of other economies. The results validate the view that Brexit will decrease domestic demand in the UK, thus hitting Britain's trading partners. The analysis also shows that exchange rate shocks in the UK have a significant negative impact on stock market indices worldwide.

Warsaw, Poland Arkadiusz Michał Kowalski

Contents

Notes on Contributors

Andreas Bielig is a member of scientific staff of the German Economy Research Department of the World Economy Research Institute in the Warsaw School of Economics. In 2006–2012 he established as first DAAD Fachlektor Wirtschaft with a DAAD scholarship the institution Fachlektorat Wirtschaft of Polish-German Academic Forum at the Warsaw School of Economics. He graduated in Economics from the University of Leipzig (Dipl.-Vw.), gained his Ph.D. degree in Economics (Dr. rer. pol.) at the Technical University of Ilmenau. Habilitation in Economic Sciences was granted him at Warsaw School of Economics. He worked as research fellow in area of environmental economics and sustainability at the Helmholtz Research Institute (UFZ) in Leipzig. He is also senior research fellow of Max Planck Institute for Innovation and Competition in Munich in fields of innovation and intellectual property. His scientific expertise includes: Intellectual Property, Economic Analysis of Law, Institutional Economics, International Economics, Economic Policy, German Economy and German-Polish Economic Relations, Automotive, Sustainability.

Anna Maria Dzienis, Ph.D. is employed as Adjunct Professor at the Department of East Asian Economic Studies, World Economy Research

Institute, SGH Warsaw School of Economics. She graduated from SGH Warsaw School of Economics. Pursuing her interest in Japanese culture, she simultaneously engaged in Japanese Studies at the University of Warsaw. As a Japanese Ministry of Education scholarship holder, she became a research student at Okayama University, before enrolling in a Ph.D. course at OU's Graduate School of Humanities and Social Sciences, which she successfully finished in 2012. For several years she combined both her fields of expertise working as an economist at the Embassy of Japan to Poland. Research of Anna Maria Dzienis concentrates on the problem of interregional migrations, labour market and current socio-economic and political issues in Japan.

Krzysztof Falkowski, Ph.D. in Economics, Assistant Professor in the World Economy Research Institute at the Warsaw School of Economics (SGH). He is the author of many publications on international economic relations in Eastern Europe and in CIS countries (focusing on Eurasian Economic Union member states). His research interests focus on the issues of international competitiveness and innovativeness of economies, institutional aspects of competitiveness and economic development, as well as systemic (including institutional) transformation in Eastern European and CIS countries, the economic situation in these countries and their position in the contemporary world economy. He has been holding lectures on i.a. International competitiveness, Dimensions of Competitiveness, Development Economics, as well as the World Economy.

Günter Heiduk Professor of Economics, studied Economics at Friedrich-Alexander-University Erlangen-Nuernberg and Albertus-Magnus University Cologne, Germany, where he got his Diploma in 1968. After finishing his Ph.D. dissertation as well as his habilitation at RWTH Aachen University he had been appointed Assistant Professor. In 1981 he accepted an appointment as Professor of International Economics at Duisburg-Essen University (former Duisburg University) from where he retired in 2007. During his tenure he served as Dean of the Faculty of Business and Economics as well as Vice-Rector for International Academic Affairs. He had been invited as Visiting Professor/Researcher at Fudan University, Kyoto University, Tokyo

Keizai University, Wuhan University. He also worked as long- and short-term expert in several higher education projects of the European Commission (including as Academic Advisor to the EU-China European Studies Centre Programme, 2004–2008). Since December 2007 he had been appointed Professor at the World Economy Research Institute/East Asian Research Unit at SGH Warsaw School of Economics, Poland. With regard to research, he published 25 books and more than 75 articles ranging from theory of international trade to economic integration in Asia and Europe. His most recent research interests involve China's Belt & Road Initiative.

Arkadiusz Michał Kowalski is Associate Professor and Head of the Department of East Asian Economic Studies at the World Economy Research Institute in the Warsaw School of Economics, Poland. His research and academic teaching focuses on innovation systems, clusters, international competitiveness, internationalisation of firms and economies in East and South-East Asia. He has been involved as a manager or researcher in different European or domestic research projects in these areas, which has resulted in more than 60 publications, including books, chapters, articles in scientific journals and expert reports. He has been also engaged into consultancy work with international organisations, governmental bodies, national parliament, enterprises and clusters. Other experiences include visiting scholarships at the University of Southampton (2004–2005), University of Glasgow (2005–2006), University of Richmond, USA (2011–2012) and Chulalongkorn University, Bangkok (2016–2017). He is a member of the East Asian Economic Association, of the Association of Polish Economists and of the Microeconomics of Competitiveness (MOC) Network, coordinated by the Institute for Strategy and Competitiveness, Harvard Business School, Boston, USA.

Marta Mackiewicz, Ph.D. is an Assistant Professor at the Department of East Asian Economic Studies at the World Economic Research Institute, Warsaw School of Economics. She is also a director of a Policy and Research Team at Ecorys Polska. She deals with economic research for 18 years. She is experienced in coordination, supervision and implementation of research and consultancy projects in the fields of competitiveness and innovation, as well as evaluation projects for a wide range of regional, national and international institutions. She is an author and co-author of publications and reports focusing on innovation issues.

Agnieszka McCaleb, Ph.D. is an Assistant Professor at the Department of East Asian Economic Studies at the World Economy Research Institute of the Warsaw School of Economics. She holds a Ph.D. in Economics from the Warsaw School of Economics and a M.A. in Sinology (Chinese Studies) from the Warsaw University. In 2016, she was awarded the Dekaban-Liddle Scholarship at the Adam Smith Business School of the University of Glasgow. Agnieszka's research interests include Chinese multinational corporations, foreign direct investment, China's national innovation system and competitiveness.

Tomasz M. Napiórkowski, Ph.D. is an Assistant Professor at the Warsaw School of Economics in Poland and a Lecturer at the Łazarski University. Since he returned to Poland in 2011 from the USA (where he studied and worked at the Old Dominion University), he authored/co-authored various scientific texts (published in both, in Polish and English), conducted a series scientific projects (including grants from the Polish National Science Centre, INSO 3 for the European Commission, DG Enterprise) and continues to implement economic theory through consultancy projects (for Master Card, The Conference of Financial Companies in Poland and Krajowy Rejestr Długów). Tomasz focuses on macroeconomic conditions, especially Foreign Direct Investment, as well as on forward-looking topics like innovation's role in economic growth.

Józef Olszyński is Associate Professor and the Head of the German Economy Research Department of the World Economy Research Institute in the Warsaw School of Economics, and Director of the Polish-German Academic Forum at the Warsaw School of Economics. He started his employment at the Warsaw School of Economics in 1972, with 16 years break due to work in the Polish diplomatic service in Germany. In years 1993–2000 and 2003–2007, he managed the Economic and Commercial Department of the Embassy of Poland in Germany. He completed Masters degree in Economics in 1972. In 1978, he completed his Ph.D. in Economics. In 1990, habilitation in international economic relations was granted to him.

Piotr Ostaszewski Professor at the Department of East Asian Economic Studies at the World Economy Research Institute, SGH Warsaw School of Economics, currently on sabbatical. Present position: the Ambassador of the Republic of Poland to the Republic of Korea.

Mariusz-Jan Radło Associate Professor of economics and Head of the Global Economic Interdependence Department at the Warsaw School of Economics—with experience in private, central government and academic sectors. Professor Radło has participated in many research and consulting projects concerning economic policy, strategy development as well as sector and macroeconomic issues. He was a visiting professor and lecturer—among others—at the Aarhus School of Business in Denmark, the Wirtschaftsuniversität Wien in Austria and Corvinus University of Budapest in Hungary.

Ewa Sońta-Drączkowska, Ph.D. is researcher and lecturer at the Warsaw School of Economics, and project management practitioner with over 18 years of experience in implementing projects both in the corporate environment and in technology startups. She has worked in various project roles ranging from business analyst to project manager and international programs. She holds the following certificates: PMP, Prince2 Practitioner, Managing Successful Programs [MSP] Practitioner, Agile PM, Design Thinking for Education—Hasso Plattner Intitute. She is a member of the Polish-German Academic Forum, member of the Project Management Institute and EURAM—the European Academy of Management, which brings together leading researchers in the field of management science.

Anna Sznajderska, Ph.D. works as researcher at the Department of East Asian Economic Studies at Warsaw School of Economics. She is Senior Economist at the Economic Institute at the National Bank of Poland, where she is responsible for modelling monetary transmission mechanism. She earned her Ph.D. from Warsaw School of Economics in 2014. She has master degree both in Mathematics (2009, University of Warsaw) and in Economics (2010, University of Warsaw). Her research interests include monetary policy, econometric modelling and international macroeconomics.

Jürgen Wandel is an Associate Professor at the World Economy Research Institute and Polish-German Academic Forum of the SGH Warsaw School of Economics. He has working experience as senior researcher and university lecturer in economics in Germany, CIS countries and Poland as well as representative of the German Academic Exchange Service (DAAD) in Kazakhstan. His research interests are institutional change and economic development in emerging markets, in particular post-Soviet countries as well as in Germany, Polish-German economic relations, economic systems, competition theory and policy, Institutional Economics, Public Choice and the Austrian School of Economic Thought. His teaching includes among others Microeconomics, European Integration, International Economics, History of Economic Thought and Economic Ethics.

Marzenna Anna Weresa Professor of Economics, since 2005 director of the World Economy Research Institute; in 2016 elected as Dean of the Collegium of World Economy at the Warsaw School of Economics for the period of 2016–2020. She holds a Ph.D. degree in Economics (1995) and a habilitation degree (D.Sc.) in Economics (2002). In 1999–2000 she worked as a research fellow at the University College London, UK. Her research and academic teaching focus on international economics, and economics of innovation in particular, issues relating to FDI, technology transfer, innovation systems as well as the effects of FDI and foreign trade on competitiveness. She authored and co-authored over 100 books and scientific articles. She has carried out many advisory projects for enterprises and governmental organisations in the field of internationalisation strategies, R&D and innovation. Since 2012, she works as an expert of the European Commission providing advice on policies for research and innovation. In 2012–2015, she was a member of the High-Level Economic Policy Expert Groups ("Innovation for Growth—I4G" and "RISE"). In 2017, she joined the European Commission expert group "Economic and Societal Impact of Research and innovation (ESIR)" providing economic analyses and recommendations in view of supporting policy implementation. She was also appointed chair of the EC expert panel working on "Mutual Learning Exercise—The evaluation of business R&D grant schemes."

List of Figures

List of Tables

1

Theoretical Aspects of Regional Disintegration and Its Consequences for International Competitiveness

Arkadiusz Michał Kowalski

1.1 Introduction

The theory of regional economic integration has developed extensively in recent decades in response to growing integration processes in Europe and elsewhere. The 23 June 2016 referendum in Britain showed, however, that regional integration does not have to be a one-way process as it was once thought to be. While there are well-developed models of regional integration in economic theory, there is little in the way of analytical explanation of the mechanics of disintegration. This is because integration was for many years commonly viewed as a beneficial process, while disintegration was seen as undesirable, which led to a normative bias in research on regional integration. This gap is the main rationale to formulate the aim of this chapter, which is to contribute to developing the theory of regional economic disintegration and gauging

A. M. Kowalski (✉)
Collegium of World Economy,
SGH Warsaw School of Economics, Warsaw, Poland
e-mail: arkadiusz.kowalski@sgh.waw.pl

A. M. Kowalski (ed.), *Brexit and the Consequences for International Competitiveness*, https://doi.org/10.1007/978-3-030-03245-6_1

its consequences for the international competitiveness of economies in areas such as international trade and the mobility of capital. It must be noted that as the EU–UK negotiations on Brexit started in 2017 and the rules and exact timetable of Britain's exit from the European Union were not yet known, different scenarios were possible, including an option referred to as a "soft Brexit", implying Britain's continued close ties with continental Europe in areas such as trade, investment and migration.

1.2 Regional Integration vs. Regional Disintegration

In recent decades, regional economic integration—defined by Balassa (1961) as "the abolition of discrimination within an area", and by Kahnert et al. (1969) as "the process of removing progressively those discriminations which occur at national borders"—has been a notable trend in the global economy. It involves the establishment of commonly accepted transnational rules on economic activity that lead to greater trade and cooperation between countries. The major examples of economic integration in the global economy, such as the European Union, the Association of Southeast Asian Nations (ASEAN) and the North American Free Trade Agreement (NAFTA), show that a key factor in this process is geographic proximity. In many cases, neighbouring countries become involved in integrative activities because of factors such as relatively short transportation distances, similar consumer tastes and needs, fairly established distribution channels, common history and an awareness of common interests. However, this is not always the case, as similar consumer tastes, for example, cannot be pointed out as a reason for regional integration between the USA and Mexico as part of the North American Free Trade Agreement (NAFTA).

There are two approaches in which we can analyse regional economic integration:

- as a continuing, step-by-step process (dynamic approach) whereby boundaries between national states become less discontinuous, thus leading to the formation of a more comprehensive system (Mennis and Sauvant 1976);
- as a state of affairs (static approach) representing the present level of integration between national economies, which may take on the form first pointed out by Balassa (1961): a free-trade area (FTA), a customs union (CU), a common market (CM), an economic union and complete integration.

In a dynamic perspective, Brexit shows that regional economic integration is not a one-way process and that it may be reversed and turned into regional economic disintegration. Hence, the traditional view of integration as a process where countries deepened cooperation and subsequently switched to modes involving stronger commitment—starting from a FTA, through a customs union, a common market, an economic union, to complete integration—is being challenged. The characteristics of different levels of regional integration, a process that may also turn into disintegration, are presented in Table 1.1.

As the level of economic integration increases so does the complexity of the process involving a set of numerous regulations, enforcement and arbitration mechanisms. However, regional integration does not have to always start with a preferential trade agreement (PTA) or a FTA and end with full integration. For example, the European Union started out as a CU, whereas NAFTA will probably never go beyond the FTA stage. Economic disintegration does not have to be a simple reversal of this process. In general, a devolution of economic integration could occur if the complexity it creates comes at a cost that may undermine the competitiveness and is no longer judged to be acceptable by society.

It should be also remembered that reaching a formal agreement does not necessarily ensure real integration between member states, as exemplified by regional integration arrangements in Africa. On the one hand, identifying the process of economic integration with

Table 1.1 Different levels of regional integration and disintegration processes

Integration process	Type (level)	Principal features	Disintegration process
↓	Preferential trade agreement (PTA)	Tariffs between the members of the agreement are reduced (or eliminated) only for some goods or services, sometimes unilaterally	↑
	Free-trade area (FTA)	No internal tariffs and import quotas Each member determines its independent trade policies with all countries outside the agreement	
	Customs Union (CU)	No internal tariffs and import quotas Harmonisation of external trade policy: Establishing a common external tariff (CET) and import quotas on goods entering the region from third-party countries	
	Common market (CM)	As for customs union above Free movement of factors of production such as labour, capital, and other resources within the region	
	Economic Union	As for common market above Coordinated monetary and fiscal policies as well as labour market, regional development, transportation and industrial policies	
	Political Union	Common home and judicial policies and a common foreign and security policy	

Based on Hill (2016)

membership in the group is debatable; on the other, it is controversial to put an equality sign between disintegration and exit from an integration grouping. For example, Poland's integration with the

European Union occurred many years before the country's formal accession. Meanwhile, a situation in which a member state ignores a judgement of the Court of Justice of the European Union is a mild form of disintegration. An example of gradual decomposition to the point of full institutional disintegration was the break-up of the Council for Mutual Economic Assistance (CMEA) in 1991, as analysed by Marszałek (1993). In fact, there were many cases of regional disintegration processes in the past, even in ancient times. For example, Goldsworthy (2009) analyses the process of the disintegration of the Roman Empire, finding plenty of analogies to the current EU situation.

Different scenarios are possible for the UK's exit from the European Union, and there are opinions that the final outcome will be the so-called soft Brexit, which could involve keeping strong linkages with the EU, e.g. through some form of membership in the FTA, customs union or even European single market, to guarantee free movement of goods, capital, services and labour. This means that regional economic disintegration does not have to necessarily mean that the leaving country will be totally out of the integration levels listed in Table 1.1. However, the question is which of these levels of relations between the European Union and the UK will be established after Brexit. Possible Brexit scenarios may be classified on the basis of findings by Barrett et al. (2015), as presented in Table 1.2.

One potential solution is that Britain will stay inside the European Economic Area (EEA), which provides the free movement of persons, goods, services and capital, and whose members also include non-EU countries representing the European Free Trade Association (EFTA). However, according to some studies, remaining a member of the common market or customs union will be not possible after Brexit, for example, because such an arrangement would not respect the result of the 2016 referendum. As the final decisions have yet to be made, the theoretical analysis of the consequences of regional disintegration for the movement of goods and capital will be developed further on in this chapter.

Table 1.2 Different scenarios for regional disintegration in the Brexit case

Alternative scenarios	Selected characteristics
Membership of the European Economic Area (EEA) and the European Free Trade Area (EFTA) (Norway's model of relationship with the EU)	• Access to the EU Internal Market for goods, but no full access to the internal market for financial services • Freedom to set own external trade policy, and own VAT regime • Freedom from participation in the Schengen free-movement zone • The need to abide by the EU law in relation to the EU Internal Market, and to contribute to the EU budget
Bilateral agreements with the EU and membership of EFTA (Switzerland's model of relationship with the EU)	• No obligation to apply and/or contribute to Common Agricultural Policy (CAP), Common Fisheries Policy (CFP), and structural funds • Freedom to conclude trade agreements with third countries • No obligation to transpose EU Internal Market legislation automatically into UK law • UK goods exported to the EU would have to comply with all relevant EU standards
Membership of a Customs Union with the EU (Turkey's model of relationship with the EU)	• Partial freedom to set own external trade policy • Access to the EU Internal Market for goods without the need to comply with EU Rules of Origin for non-EU countries • No contribution to the EU budget • The right to regulate its own financial sector • Common external tariff on imports from outside the UK/EU customs union • EU product standards for goods • EU common commercial policy
Bilateral Free-Trade Agreement with the EU	• Freedom to set own external trade policy and VAT regime, and to conclude FTAs with third countries • No obligation to contribute to the EU budget • UK goods exported to the EU would have to comply with all relevant EU standards

(continued)

Table 1.2 (continued)

Alternative scenarios	Selected characteristics
No preferential trade agreement with the EU	• National competence over trade policy and border control • Removal of the requirement to contribute to the EU budget and of all EU legislative rights • Most favoured nation tariffs will be applied in line with membership of the World Trade Organisation

Based on Barrett et al. (2015, pp. 1, 70–71)

1.3 Explaining Regional Disintegration with the Concept of Neo-functionalism

The so-called neo-functionalist approach to conceptualising regional disintegration was proposed by Schmitter and Lefkofridi (2016). Originally, neo-functionalism was developed by, e.g. Haas (1964) and Schmitter (1970) as an important theory of European integration, assuming higher efficiency (functionality) of regional integration in relation to actions taken by individual countries. According to the neo-functionalist approach, regional integration is a relatively steady process, involving two parallel elements: market integration and the delegation of policy-making competence to an organisation above the national level. At the core of neo-functionalism is the concept of spill overs, which refers to situations when a certain sector is placed under the authority of a central institution (such as the European Commission) and when pressures are created to extend the authority of this institution into neighbouring areas of policy, such as taxation, wages or currency exchange rates (Tranholm-Mikkelsen 1991). The theory was optimistic about regional integration, and even crises were perceived as catalysts for positive change as they led to stronger regulatory expansion. This assumption was challenged during the global financial crisis that started in 2008. It showed that European integration was not functional because EU institutions were unable to effectively deal with this crisis (Grosse 2016).

One of the proponents of neo-functionalism, E. Haas (1968, p. 16), defined regional integration as "the process whereby political actors in several distinct national settings are persuaded to shift their loyalties, expectations and political activities toward a new centre, whose institutions possess or demand jurisdiction over the pre-existing national states". From that perspective, disintegration does not have to be the same as a reversed process of integration, as authority may not be transferred back to national states but can be shifted instead to regional authorities. Thus, disintegration is not necessarily a choice between states and a central institution, as regions may emerge as important actors.

Regional disintegration was conceptualised in the neo-functionalist approach by Schmitter and Lefkofridi (2016), who state that:

- if the benefits of integration are not evenly distributed across member states and its societies, the risks of disintegration increase;
- as neo-functionalism assigns a key role to experts, those in supranational institutions and national parliaments, disintegrative forces can come about if there is heterogeneity in preferences between member states and regional institutions;
- disintegrative pressures emerge if the process of regional integration is not gradual and requires "a radically different mode of decision-making or conflict resolution" and if member states fail to implement EU decisions;
- regional integration unavoidably generates conflict among member states. If the conflict is too big to be effectively handled by a central institution, it will mobilise an increasingly wider public expressing a greater diversity of opinions.

1.4 New Intergovernmentalism and the Implications for Regional Integration and Disintegration Processes

As neo-functionalists stressed the relevance of Community bodies in the process of regional integration, the "new intergovernmentalism" is an alternative theory that highlights the continuing importance of nation states (Hoffmann 1995, pp. 71–106). Being sceptical about the "community method" as the main modus operandi of the European Union, the theory postulates an active role for member states in advancing stronger cooperation in areas in which they have competence. Hence, in areas where there was no EU competence, integration can be advanced only by the member states.

The new intergovernmentalism was used as a theoretical explanation of the process of regional disintegration by Bickerton et al. (2015), who formulated the hypothesis that the European Union is in a "state of disequilibrium" as there are constant tensions between member states. However, even from this perspective, this concept is not used for predicting a potential break-up of the European Union, but it rather explains a particular feature of European integration since Maastricht. On the other hand, according to Fabbrini and Puetter (2016, p. 488), "if consensus is threatened or impossible constitutional or redistributive adjustments are either inevitable in order to mitigate tensions and asymmetries between the Union's member states or, if these options are not available, there is a risk of disintegration". Hence, the new intergovernmentalism does not assume automatic regional disintegration, but has clear disintegrative elements, and may be used to explain how the disintegration process occurs.

1.5 The Concept of International Competitiveness

When analysing the consequences of regional economic disintegration for international competitiveness, it is necessary to introduce the concept of economic competitiveness. Although it is one of the most widely used terms in modern economics, there is a significant lack of consensus on what it really means. Ketels (2015) calls for a shared definition of the term in order to make it a useful category for policy dialogue, proposing the adoption of the Aiginger and Vogel (2015) definition of competitiveness as the "ability of a country (region, location) to deliver the beyond-GDP goals for its citizens". This definition reflects the comprehensive nature of the concept of competitiveness, which encompasses different dimensions, in terms of both different types of economic performance and the geographical perspective. With respect to beyond-GDP objectives underlined in the definition, competitiveness refers to not only income levels, but also other perspectives, including social, ecological and institutional. Under the methodology used by the Warsaw School of Economics' World Economy Research Institute in its annual competitiveness reports (e.g. Weresa and Kowalski 2018), competitiveness is understood as an economy's ability to achieve:

1. a sustainable increase in the standard of living (income competitiveness),
2. an improvement in a country's position in the global marketplace (trade competitiveness),
3. enhanced investment attractiveness, mostly for foreign capital (investment competitiveness).

In general, the concept of competitive ability is distinguished from that of competitive position in research reports. Competitive ability is also called factorial competitiveness, as it is assessed on the basis of a number of factors that describe the size, structure and use of productive resources, the socio-economic system, the government's economic policy and the international economic environment. All these factors determine the ability to compete in foreign markets and to achieve a certain competitive position. Meanwhile, the competitive position indicates the level of economic development achieved by a country and reflected in the level of income, as well as the efficiency with which factors of production are used, and the country's position in foreign trade (Kowalski 2013, p. 77). Meanwhile and Gorynia (1998, p. 35) proposes a division into an *ex post* competitive position, i.e. the current competitive position, which is the result of the implemented competitive strategy, and an *ex ante* competitive position, understood as a future (prospective) competitive position, which is determined by the relative ability of the company (compared to the abilities of its competitors) to compete in the future, and thus representing its competitive potential.

Economic competitiveness and its determinants can be analysed at different levels. With respect to the level of aggregation or geographical dimension, competitiveness can be analysed at different system dimensions:

1. microeconomic competitiveness (single company level),
2. mesoeconomic competitiveness (regional or sectoral perspective),
3. macroeconomic competitiveness (country level),
4. megaeconomic competitiveness (group-of-countries perspective),
5. metaeconomic competitiveness (competition between different models of capitalism).

It should be noted that all the above-mentioned levels are strongly interconnected, as it is the successes of single companies that determine the prosperity of local regions, which subsequently contribute to the development of particular countries forming bigger groups of national economies sharing similar characteristics (Kowalski 2018).

1.6 The Impact of Regional Disintegration on Production and International Trade

Regional disintegration means that some effects of the customs union are eliminated. One of the most important is the trade creation effect, which occurs when domestic production of a good in a member country is replaced by imports of the same good from another member country within the customs union because of lower production costs. From this perspective, Brexit would reverse this process, meaning that some goods would be produced in the UK instead of being imported from continental Europe. However, this would not benefit customers because of higher prices for such products. In the same way, exports of British products to the European Union would be diminished because of the disintegration process.

Another economic consequence of regional disintegration is connected with the trade diversion effect, which occurs when imports from a non-member country are replaced by imports from a member state because of the application of common customs tariffs to non-member countries. In this scenario, part of Britain's exports to different EU countries may be replaced by trade between member states.

Regional disintegration may lead to some dynamic effects. Based on the results of a systematic investigation of the dynamic effects of economic integration that was first carried out by Balassa (1961), it is possible to identify the following effects of regional disintegration:

- adverse effects on economies of scale: as globalisation and regional integration have brought with them a fragmentation of production and vertical specialisation, leading to economies of scale, the regional disintegration process will diminish the disintegration of production, at least to some extent, limiting the economies of scale experienced by an economy leaving the union;
- adverse effects on competition: leaving the customs union diminishes the market in comparison with free trade and reduces the level of competition, with negative consequences for efficiency;
- adverse effects on capital formation and investment: leaving the customs union may diminish outside investment. This would have negative effects on international competitiveness as investment is usually attracted to the most productive and competitive sectors;
- adverse effects on the terms of trade in the economy leaving the union; the terms of trade may deteriorate as the country will have less bargaining power;
- negative impact on technological progress and innovation, as undertaking joint research and development (R&D) or international cooperation for the sake of technology transfer are becoming the key drivers of the innovativeness of economies, especially in the context of internationalisation processes involving innovation (so-called techno-globalism).

1.7 The Impact of Regional Disintegration on the Mobility of Capital: Foreign Direct Investment

Regional economic disintegration, which increases regional trade barriers and investment restrictions, negatively impacts different forms of firm internationalisation processes, including foreign direct investment (FDI) flows. FDI, one of the possible channels of international economic involvement, represents a large part of the increasing and all-encompassing trends towards globalisation. Basically, it is a part of multinational companies' attempts to overcome obstructions to foreign trade, licensing, joint ventures, management contracts and so on, so the reason for its growth at the global level is the imperfections in the world economy and protective trade policies pursued by different countries. According to the definition by the International Monetary Fund (2009, p. 100), foreign direct investment "is a category of cross-border investment associated with a resident in one economy having control or a significant degree of influence on the management of an enterprise that is resident in another economy".

The inflow of foreign direct investment results in many benefits for the host economy, so different countries compete trying to offer better conditions and incentives to attract multinational companies. For example, multinational companies can bring new technology and provide technical assistance, which is especially valuable for developing countries. Foreign direct investment generates jobs for both skilled and unskilled labour and contributes to GDP growth. These benefits may be reduced by regional disintegration, which diminishes the capital inflow to the economy leaving the integration grouping. One of the main reasons is smaller market size because of regional disintegration. This may motivate foreign companies to move their investment from a country leaving an integrated economic area (i.e. the UK in the case of Brexit) to other countries within this area. According to different studies, the size of the host market is an important determinant of attractiveness to FDI (Globerman and Shapiro 2003). This especially applies to market-seeking investors, whose motivation is focused on gaining access to particular markets through local production and distribution, rather than by exporting from the home country or from a third country. Entering a new market provides a company with a chance to achieve economies of scale and to be more competitive. Market-seeking investment is attracted by factors including host country market size, per capita income and market growth.

Regional disintegration may provide national economies with additional location-specific advantages that serve to attract FDI. One of the tools that may be applied when analysing this problem is the diamond model of competitive advantage proposed by M. E. Porter (1980). The model distinguishes between different location-specific advantages:

1. factor conditions,
2. demand conditions,
3. related and supporting industries,
4. industry structure and rivalry.

Regional economic integration may influence changes in location-specific competitive advantages. For example, new market boundaries can reduce the rivalry among competitors (which is the fourth component of Porter's diamond model). A smaller market will also have a negative effect on the buying power, while the bargaining power of buyers will decrease due to smaller supply (so the second component of Porter's diamond model is affected). It should be noted, however, that theoretically, the loss of an economy leaving the union depends on the size and strength of this economy. Based on the J. Dunning (1997) observation that Regional Integration Agreements may modify firm-specific advantages—which, in turn, have an impact on incentives for companies to undertake FDI—regional disintegration may weaken the geographical concentration of specific industrial activities. This is because businesses in the country leaving the union have less of an incentive to concentrate their production in the market with higher production costs, as the economies of scale are diminished. This finding is of particular relevance to industries that exhibit significant agglomeration economies, meaning benefits that companies obtain when locating near each other (Blomström et al. 1998, p. 5).

1.8 Conclusions

The Brexit referendum has changed the perspectives of European integration, which was previously perceived as a constant process of deepening cooperation among countries. Although there are some theoretical concepts that form the background for analysing regional disintegration, such as "new intergovernmentalism" and neo-functionalism, economic theory has failed to provide a deeper analytical explanation of the mechanics of this process. In the case of Britain's exit from the European Union, it must be noted that negotiations on the rules and exact timetable of the process started in 2017 and different scenarios are possible, including different variants of a so-called soft Brexit, which would imply Britain continuing to enjoy some form of a common market or customs union with continental Europe. If, on the other hand, a "hard Brexit" scenario were to pan out, removing the customs union between the UK and the EU would eliminate the trade creation and trade diversion effects. It would also lead to adverse dynamic effects on economies of scale, competition, capital formation and investment as well as terms of trade, technological progress and innovation. Disintegration processes, by increasing regional trade barriers and investment restrictions, would negatively impact different forms of firm internationalisation, including FDI flows to the UK. This would reduce many of the benefits for the British economy in areas such as job creation and technology transfer.

References

Aiginger, K., & Vogel, J. (2015). Competitiveness: From a Misleading Concept to a Strategy Supporting Beyond-GDP Goals. *Competitiveness Review, 25*(5), 497–523.

Balassa, B. (1961). *The Theory of Economic Integration*. Homewood, IL: Richard D. Irwin.

Barrett, A., Bergin, A., FitzGerald, J., Lambert, D., McCoy, D., Morgenroth, E., et al. (2015). *Scoping the Possible Economic Implications of Brexit on Ireland* (Research Series No. 48). Dublin: Economic and Social Research Institute.

Bickerton, C. J., Hodson, D., & Puetter, U. (2015). The New Intergovernmentalism: European Integration in the Post-Maastricht Era. *JCMS: Journal of Common Market Studies, 53*(4), 703–722.

Blomström, M., Kokko, A., & Globerman, S. (1998). *Regional Economic Integration and Foreign Direct Investment: The North American Experience* (Working Paper Series in Economics and Finance No. 269).

Dunning, J. H. (1997). The European Internal Market Programme and Inbound Foreign Direct Investment, Part I. *Journal of Common Market Studies, 35*(1), 1–30.

Fabbrini, S., & Puetter, A. (2016). Integration Without Supranationalisation: Studying the Lead Roles of the European Council and the Council in Post-Lisbon EU Politics. *Journal of European Integration, 38*(5), 481–495.

Globerman, S., & Shapiro, D. M. (2003). Governance Infrastructure and US Foreign Direct Investment. *Journal of International Business Studies, 4*(1), 19–39.

Goldsworthy, A. (2009). *The Fall of the West: The Death of the Roman Superpower*. London: Phoenix.

Gorynia, M. (1998). *Zachowania przedsiębiorstw w okresie transformacji. Mikroekonomia podejścia*. Poznań: Akademia Ekonomiczna w Poznaniu.

Grosse, T. G. (2016). Assumptions of the Theory of Regional Disintegration: Suggestions for Further Research. *Przegląd Europejski, 4*(42), 10–22.

Haas, E. B. (1964). *Beyond the Nation State: Functionalism and International Organisation*. Stanford, CA: Stanford University Press.

Haas, E. B. (1968). *Collective Security and the Future International System*. Denver: University of Colorado.

Hill, C. W. L. (2016). *International Business*. Irwin: McGraw-Hill Education.

Hoffmann, S. (1995). *The European Sisyphus: Essays on Europe, 1964–1994*. Boulder, CO: Westview Press.

International Monetary Fund. (2009). *Balance of Payments and International Investment Position Manual* (6th ed.). Washington, DC: International Monetary Fund.

Kahnert, F., Richards, P., Stoutjesdijk, E., & Thomopoulos, P. (1969). *Economic Integration Among Developing Countries*. Paris: Development Center of the Organization for Economic Co-operation and Development (OECD).

Ketels, C. (2015). *Competitiveness and Clusters: Implications for a New European Growth Strategy* (WWWforEurope Working Paper No. 84). Vienna: WIFO.

Kowalski, A. M. (2013). *Znaczenie klastrów dla innowacyjności gospodarki w Polsce*. Warszawa: Szkoła Główna Handlowa w Warszawie – Oficyna Wydawnicza.
Kowalski, A. M. (2018). Benefits of Broadening the Analysis of International Competitiveness: The Case of CEE Countries. *AIB Insights, 18*(1), 7–11.
Marszałek, A. (1993). *Planowanie i rynek w RWPG*. Geneza niepowodzenia, *Acta Universitatis Lodziensis. Folia Oeconomica*, Nr 118/1993.
Mennis, B., & Sauvant, K. P. (1976). *Emerging Forms of Transnational Community*. Lexington: Lexington Books.
Porter, M. E. (1980). *Competitive Strategy: Techniques for Analyzing Industries and Competitors*. New York: The Free Press.
Schmitter, P. C. (1970). A Revised Theory of Regional Integration. *International Organisation, 24*(4), 836–868.
Schmitter, P. C., & Lefkofridi, Z. (2016). Neo-functionalism as a Theory of Disintegration. *Chinese Political Science Review, 1*(1), 1–29.
Tranholm-Mikkelsen, J. (1991). Neo-functionalism: Obstinate or Obsolete? A Reappraisal in the Light of the New Dynamism of the EC. *Millennium—Journal of International Studies, 20*(1), 1–22.
Weresa, M. A., & Kowalski, A. K. (2018). Preface. In M. A. Weresa, A. K. Kowalski (Eds.), *Poland: Competitiveness Report 2018—The Role of Cities in Creating Competitive Advantages* (pp. 7–9). Warsaw: Warsaw School of Economics—Publishing.

2

Brexit and Innovation: Focus on Research and Development in the UK

Marzenna Anna Weresa

2.1 Introduction

In June 2016, the British people voted to leave the European Union. The Brexit decision has prompted a lot of discussion on its implications for innovation in the UK and the EU.

The main objective of this chapter is to identify how the UK's position will change after Brexit in terms of internationalisation of the UK's national innovation system, in particular when it comes to the country's participation in EU-funded research and innovation programmes.

According to the European Innovation Scoreboard 2017, the UK is among innovation leaders and ranks fifth in the EU in terms of innovation performance measured by the Summary Innovation Index, after Sweden, Denmark, Finland and the Netherlands (European Commission 2017b, p. 6). The UK has the advantage of having a strong

M. A. Weresa (✉)
Collegium of World Economy,
SGH Warsaw School of Economics, Warsaw, Poland
e-mail: marzenna.weresa@sgh.waw.pl

A. M. Kowalski (ed.), *Brexit and the Consequences for International Competitiveness*, https://doi.org/10.1007/978-3-030-03245-6_2

science base, and since 2010 the country's innovation performance—relative to that of the EU average measured by the Summary Innovation Index—has increased by 11.7% (European Commission 2017b, p. 66). As an EU member, the UK benefits from being involved in European research projects financed through Framework Programmes. For example, Britain received more than 16% of funding under the EU's Horizon 2020 programme over three years, the second largest proportion among EU member states after Germany's 19% (European Commission 2017a, p. 66). The UK also has the largest number of participants in the Horizon 2020 programme. In the 2014–2016 period, UK participants coordinated one in every five H2020 projects (European Commission 2017a, p. 64). Moreover, European Research Council data show that the UK in 2017 ranked first in terms of hosting successful Starting Grant applicants among early career researchers (ERC 2017).

All these facts show that the EU is an important source of R&D funding for the UK. At the same time, the UK plays a key role in EU research because it is among the top five collaboration partners for each of the remaining 27 member states (European Commission 2017b).

It should also be noted that the UK is strongly engaged in the mobility of academics and students. More than 16% of academic staff at UK universities come from other EU countries (UCEA 2017, p. 35) and about 6% of the students in UK's higher education are from other EU member states (UKCISA 2017).

These close ties between the innovation systems of the UK and other member states prompt a number of questions about Brexit and its impact on the UK's research and innovation performance:

- To what extent is Britain's national innovation system (NIS) internationalised compared to other EU member states, and what role do other EU countries play in this process?
- What is the role of the EU in research funding in the UK?
- To what extent will Brexit affect the UK's access to EU funds?
- How will the UK's position change in terms of participation in EU-funded research and innovation programmes, such as Horizon 2020, after the country leaves the European Union in March 2019?

These research questions refer to the empirical part of the study, which is preceded by a brief literature review aimed at identifying key factors determining the internationalisation of national innovation systems. The hypothesis investigated in this chapter is that Brexit will weaken the UK's research and innovation performance.

The chapter is organised as follows. The next section provides a brief literature review on the internationalisation of national innovation systems. This is followed by an assessment of the internationalisation of the UK innovation system. In particular, research and innovation collaboration between the UK and the rest of the EU are analysed. Moreover, the UK's position in the absorption of EU funds for research and innovation is analysed in depth. On this basis, some scenarios are drawn up for research collaboration between the UK and the EU post-Brexit.

2.2 Internationalisation of the UK Innovation System: The Role of Other EU Countries

According to the theory, economic integration processes produce an increase in economic cooperation among partners within an integration group (see, for instance, Baldwin 2015; Barcz et al. 2016). This also holds true for collaboration in research and development (R&D). Increased R&D collaboration within an integration bloc may contribute to the creation of new knowledge and more innovation.

Brexit, being an example of a disintegration process, will probably have the opposite effects. Closing the borders may discourage firms and individuals from continental Europe to undertake collaborative research and innovation projects in the UK. It may also limit technology transfers between the UK and the rest of the EU, thus reducing synergies and spillover effects. Two key factors shape the extent to which such disintegration effects may occur after Brexit: (1) the level of internationalisation of the UK's national innovation system and the strength of its relations with its counterparts in EU countries; (2) the result of Brexit negotiations related to research and innovation, and decisions on the extent to which short- and long-term collaboration between UK and

EU businesses, universities and research organisations will be supported from the UK and EU budgets. Therefore, the next step in our analysis is an attempt to assess the internationalisation of the UK national innovation system and to identify the role of EU funds in the UK research system.

2.3 Internationalisation of the Innovation System: A Literature Review

The internationalisation of innovation systems refers to international collaboration in research and development and the application of new knowledge as well as the international diffusion and transfer of new solutions. The internationalisation of innovation means that the value chain is spread across different geographic locations, and as a result research, innovation, production and value creation do not necessarily occur in the same country (Schwang Serger and Wise 2010). Multinational companies and their integrated strategies are one of the main drivers of this trend (Carlsson 2006). Furthermore, international cooperation in R&D and innovation as well as technology transfer and diffusion of knowledge can be supported by international research programmes as well as programmes for the international mobility of researchers, coordination of education systems and introduction of open access to research results.

Research on the internationalisation of national innovation systems (NIS) has been conducted since the early 1990s. As pointed out in the literature, interactions between national innovation systems in different countries include different forms of international transfer and diffusion of innovation (see, for instance, Niosi and Bellon 1994; Archibugi and Iammarino 2000; Weresa 2014). International science, technology and innovation (STI) activities can be divided into different groups broken down by the type of activities performed by different actors. Olbrich and Witjes (2014), in their study on the internationalisation of the national innovation system of South Korea, focus on green technologies and propose a structured catalogue of international STI activities (Table 2.1).

Table 2.1 Catalogue of international STI activities

Type of activity	Actors and specification of their activities
Research and development	**Public and private organisations** (universities, research organisations, enterprises, etc.) involved in the international production of knowledge (basic and applied research conducted abroad or domestically by foreign researchers)
Education	**Universities** as main actors conducting topic-specific study and exchange programmes and their strategic content orientation
Introduction and modification of organisations	**Policy makers** that create new research organisations, agencies, coordination committees or ministries for the purpose of facilitating and engaging in international STI activities
Assistance and consultancy	**Government actors, foundations or other non-governmental organisations** that finance R&D contributing to long-term cooperation, and providing starting capital or consultancy services supporting development of international linkages
International networking	**Public and private organisations** (universities, research organisations, enterprises, etc.) conducting bilateral and multilateral cooperation with foreign partners aimed at integrating knowledge and facilitating its international dissemination

Source Adapted from Olbrich and Witjes (2014, p. 72)

The above is a clear theoretical division of STI activities into five distinctive categories. However, this division has some drawbacks. When analysing the roles of different actors in the internationalisation of national innovation systems, one should take into account the fact that some actors may be involved in two or more activities. Furthermore, several types of activities have been excluded from this catalogue. First, cooperation in innovation conducted by enterprises from different countries is not explicitly mentioned. It is one of the core elements of the NIS internationalisation process, but in the classification proposed by Olbrich and Witjes this kind of cooperation seems to be part of the research and development category. Second, international coordination of STI policy has not been included in the above catalogue. It might not be important globally, but it is a vital aspect of the internationalisation of national innovation systems in EU countries. EU innovation strategies—such as the Lisbon agenda (European Parliament 2000), which set out the goal of establishing a European Area of Research and

Innovation, and the Europe 2020 strategy, with its Innovation Union initiative (European Commission 2011)—are examples illustrating how the innovation policies of individual EU member states have been coordinated at the EU level. The original classification by Olbrich and Witjes was therefore expanded in this study to include two more types of international STI activities. Also, individual STI activities were defined more precisely, accompanied by a list of indicators used for a comparative analysis of the internationalisation (Europeanisation) of the UK national innovation system. Table 2.2 specifies the key facets of internationalisation of national innovation systems and related STI activities as well as the main actors involved and selected measures of NIS internationalisation (Europeanisation).

The NIS internationalisation indicators listed in Table 2.2 do not exhaust all the possible ways of measuring the process. The use of indicators in empirical studies depends on the selection of the studied sample of countries, but is limited by the availability of statistical data. The scope of this chapter limits our analysis to the basic indicators of internationalisation of the British NIS: 10 of the 17 indicators given in Table 2.2.

2.4 Innovation System Internationalisation Before Brexit: UK Compared with Selected Other EU Countries

The main conclusion from empirical research on NIS internationalisation, one related to EU research and innovation systems, was summarised by Carlsson (2006, p. 59): "The European Union appears to be the only major supranational scientific and technological block now emerging. Japan seems to be much less internationalized (and its internationalization is aimed principally at the USA), and Canada-USA interaction (in spite of NAFTA) is less evident than in the EU." Therefore, in the context of Brexit, it is worth examining how the UK research system will be affected when the country exits the European Union.

This section is aimed at assessing the internationalisation of the British NIS compared to other EU member states. Such an analysis

Table 2.2 Major facets of NIS internationalisation/Europeanisation

Type of STI activities	Main actors involved	Examples of indicators
Research and development	Universities, research organisations, enterprises, involved in international production of knowledge and its financing (basic and applied research conducted abroad or domestically by foreign researchers) and developing international cooperation aimed at integrating knowledge and disseminating research results internationally Government actors, foundations or other non-governmental organisations that finance international cooperation in R&D and provide starting capital or consultancy services supporting development of international linkages	– R&D expenditure from foreign/European sources relative to total R&D expenditure – Involvement of domestic actors in EU framework programmes (value of grants received) – R&D expenditure of foreign-owned enterprises relative to domestic business expenditure – Scientific publications among the top 10% most cited publications worldwide as % of total scientific publications of the country – Number of research publications developed in cooperation with foreign partners per capita – Co-patenting with foreign (European) partners – Licence and patent revenues from abroad as % of GDP
International/ European coordination of STI policy	Policy makers, international organisations designing and implementing transnational innovation strategies and programmes. Policy makers that create new research organisations, agencies, coordination committees or ministries aimed at facilitating international STI activities	– The state of implementation of EU research and innovation strategies (Innovation Union Initiative since 2010) – Eco-innovation index[a] – The ratio of turnover from products new to the enterprise and new to the market as a % of total turnover

(continued)

Table 2.2 (continued)

Type of STI activities	Main actors involved	Examples of indicators
Education and international/European mobility of students and researchers	Universities involved in designing and implementing foreign exchange programmes; individual researchers and students that take part in these programmes	– Participation of foreign students in tertiary education; number of foreign doctoral students in a country as a percentage of the total number of doctoral students overall and/or by sector/technology – Number of foreign researchers and doctoral students in a country relative to the number of domestic researchers and doctoral students working abroad (overall and by sector/technology) – Foreign employees in R&D sector relative to total R&D employment
International/European cooperation in and innovation (including international networking)	Enterprises, universities, research organisations involved in bilateral and multilateral innovation cooperation with foreign partners Governmental agencies, foundations or other non-governmental organisations that facilitate international transfer and diffusion of innovations	– Enterprises undertaking all kinds of innovation cooperation with foreign partners (as a percentage of the total number of innovative enterprises) – Innovative companies that received funds to finance international cooperation from public funds – High-tech imports/exports as % of total – Involvement in formal and informal innovation networks (e.g. Enterprise Europe Network, international/European cluster networks)

[a]The eco-innovation Index is based on 16 sub-indicators from eight contributors in five thematic areas: eco-innovation inputs, eco-innovation activities, eco-innovation outputs, resource efficiency outcomes and socio-economic outcomes. The overall score of an EU member state is calculated by the unweighted mean of the 16 sub-indicators. It shows how well individual member states perform in eco-innovation compared to the EU average. It is one of the indicators related to the Innovation Union strategy (Eurostat)

Source Own work

makes it possible to draw conclusions about the potential consequences of Brexit for the UK's STI sector and its development.

The empirical part of this chapter is based on statistics from the following databases: Eurostat, the European Innovation Scoreboard database, the World Intellectual Property Organisation (WIPO) and selected UK databases, such as the Office for National Statistics and the Universities and Colleges Employers Association. The analysis covers the period of 2010–2015 (or 2014 where data for 2015 is not available), but in some cases the latest year available was 2012. The comparative analysis of the internationalisation of the UK national innovation system covers two groups of countries:

1. EU innovation leaders, i.e. Sweden, Denmark, Finland and the Netherlands—peers for the UK in terms of innovation performance as measured in the European Innovation Scoreboard 2017 (European Commission 2017b, p. 6);
2. Germany and France—worse innovation performers than the UK, ranked sixth and eleventh in the EU (European Commission 2017b, p. 6), respectively, but these two countries are the UK's peers given their size measured by the GDP and population.

Ten indicators of NIS internationalisation have been considered in this comparative analysis, at least one for each of the facets of internationalisation presented in Table 2.2.

Figure 2.1 compares the internationalisation of the British national innovation system with other innovation leaders in the EU. All these countries in 2016 performed better than the UK in terms of NIS. However, when it comes to the internationalisation of NIS, the UK tops this group of countries on four of ten considered indicators. The UK leads the group with regard to the share of EU funds in R&D expenditure as well as in funding research under the Seventh Framework Programme (FP7). Another relative strength of the UK is in the internationalisation of its education system. The UK has the highest participation of foreign students in tertiary education and in doctoral programmes among the analysed countries (Fig. 2.1).

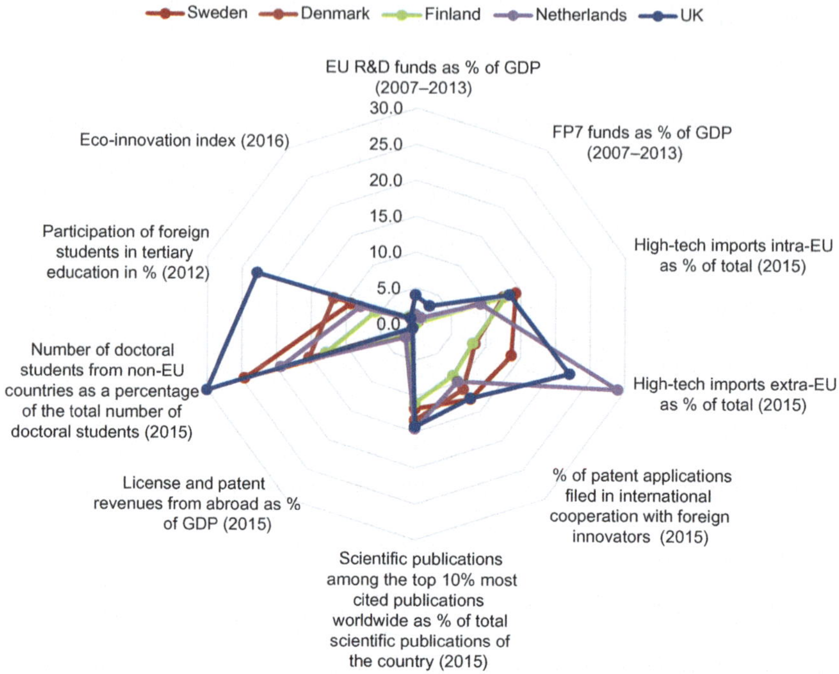

Fig. 2.1 Internationalisation of NIS: the UK and other innovation leaders compared (*Source* Own elaboration based on data from Eurostat, WIPO and Royal Society [2015])

Three other indicators of NIS internationalisation, i.e. the percentage of scientific publications among the top 10% most cited publications worldwide, co-patenting with foreign innovators, and the share of high-technology imports, have also been quite high in the UK, ranking this country second in the analysed group (Fig. 2.1).

The UK is one of the worst performers in the analysed group in terms of two indicators: licence and patent revenues from abroad as a percentage of GDP, and the eco-innovation index (Fig. 2.1). The above NIS internationalisation analysis confirms that the UK innovation system has achieved a level of internationalisation similar to those exhibited by other innovation leaders in Europe.

A comparison of Britain with Germany and France in terms of NIS internationalisation shows that the UK outperforms these two countries in six of ten indicators (Fig. 2.2).

Germany surpassed the UK in the absorption of R&D funds from the EU budget. However, the UK leads the way if structural funds are excluded and only FP7 grants are counted. Germany is also slightly better in implementing eco-innovation (Fig. 2.2).

France is better than the UK in attracting foreign doctoral students. In 2015, doctoral students from non-EU countries made up 33% of the total number of doctoral students in France, while the proportion in the

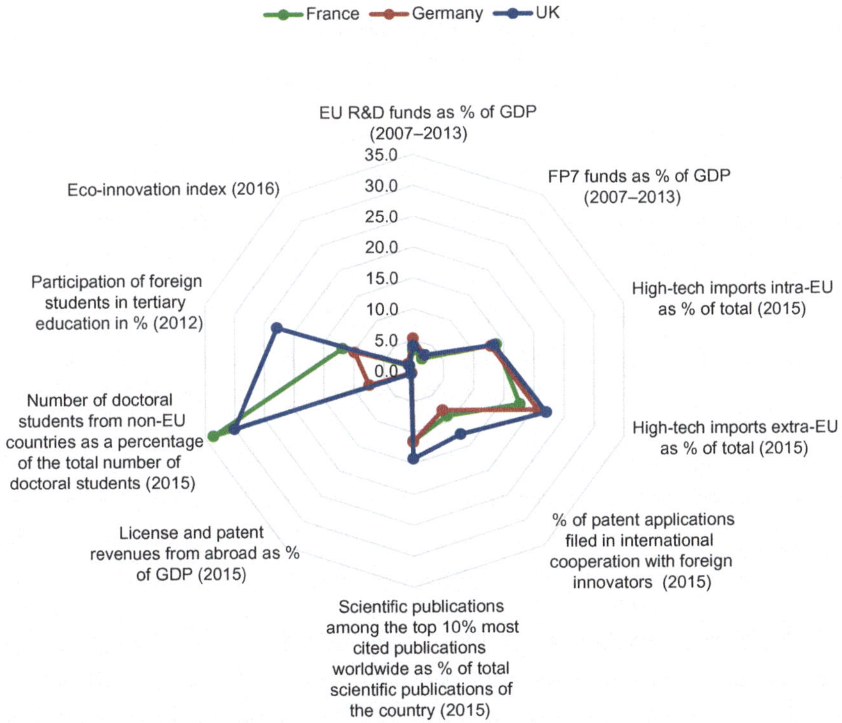

Fig. 2.2 Internationalisation of NIS: the UK compared to Germany and France (*Source* Own elaboration based on data from Eurostat, WIPO and Royal Society [2015])

UK was 30%. Another relative weakness of the UK compared to France is the share of non-EU countries in total high-tech imports, although the difference was insubstantial in 2015: 13.8% in France versus 13.4% in the UK. In Germany, high-tech goods from non-EU countries represented 12.9% of total imports (Fig. 2.2).

To sum up, the overall performance of the UK in terms of NIS internationalisation is among the best in the two analysed groups of EU countries, as the UK occupies either first or second place in both country groups for most internationalisation indicators (see Figs. 2.1 and 2.2).

2.5 Research and Development: The Role of EU Funds in the UK

Science and innovation contribute to a country's prosperity and well-being. However, nowadays more emphasis is being placed on the benefits to society and contribution to widely shared societal goals (Long and Blok 2017, p. 64). Public and corporate R&D expenditure, education and training as well as infrastructure and regulations support innovative activities. Investment in R&D is regarded as one of the basic components of a strong national innovation and technology base. Therefore, it is worth analysing R&D investment in the UK and predicting how it will change in the post-Brexit period.

The British science and research system is strongly outward-oriented, as reflected by the high share of R&D funding from abroad. From 2010 to 2015, the share of foreign funds in total R&D expenditure in the UK hovered around 17%. It was much higher than the EU average, which stood at 10% in 2014. Comparing the UK's position in terms of foreign participation in R&D funding with those of other EU countries, in particular the country's peers in Western Europe, it should be noted that the share of foreign sources of financing R&D in the UK is similar to those in Ireland, Austria, and Finland, and three times as high as in Germany and Sweden (Table 2.3). This shows that the UK is more dependent on foreign financing for its STI sector than other countries.

Table 2.3 R&D expenditure from foreign sources relative to domestic expenditure in 2010–2015 in the UK and other EU countries (%)

GEO/TIME	2010	2011	2012	2013	2014	2015
European Union (28 countries)	**8.9**	**9.2**	**9.7**	**9.9**	**10.0**	–
Euro area (19 countries)	7.3	7.5	7.9	8.4	8.5	–
Belgium	13.3	13.0	13.0	13.2	–	–
Bulgaria	39.6	43.9	46.3	48.3	50.9	–
Czech Republic	13.9	19.7	25.9	27.2	30.5	32.5
Denmark	7.2	7.1	7.2	6.7	–	6.7
Germany	3.9	4.2	4.3	5.2	5.0	–
Estonia	11.4	11.9	10.0	10.3	12.5	12.2
Ireland	17.0	20.3	21.0	18.6	18.6	–
Greece	11.9	14.8	15.8	14	13.2	12.8
Spain	5.7	6.7	6.6	7.4	7.4	–
France	7.5	7.7	7.6	8.0	7.8	–
Croatia	9.9	11.6	14.4	15.5	12.8	14.5
Italy	9.8	9.1	9.5	9.7	9.3	–
Cyprus	15.0	14.1	17.5	19.6	23.7	–
Latvia	33.4	51.0	50.4	51.6	44.2	45.0
Lithuania	19.9	28.4	33.2	37.1	33.8	34.6
Luxembourg	20.6	19.5	34.4	32.3	–	–
Hungary	12.4	13.5	15.4	16.6	17.5	15.0
Malta	12.2	18.4	21.7	23.3	21.0	21.3
Netherlands	–	11.3	12.5	12.2	12.7	15.1
Austria	16.1	16.9	16.1	16.6	16.1	15.9
Poland	11.8	13.4	13.3	13.1	13.4	16.7
Portugal	3.2	6.0	5.2	6.1	5.6	–
Romania	11.1	12.1	14.4	15.5	17.0	19.2
Slovenia	6.0	7.0	8.6	8.9	9.3	10.6
Slovakia	14.7	14.2	18.7	18.0	23.7	39.4
Finland	6.9	6.5	8.8	11.5	17.3	14.5
Sweden	–	11.0	–	6.7	–	–
United Kingdom	**17.6**	**17.8**	**19.8**	**18.7**	**17.5**	**17.6**

Source Eurostat data. Accessed 18 October 2017
Note "–" means that data is not available

EU support for science and research in the UK goes through the following channels (House of Lords 2016, p. 6):

- EU framework programmes (The Horizon 2020 programme in 2014–2020, and FP1–FP7 earlier); European Structural and Investment Funds (ESIF);

- Sectoral research and development programmes;
- Other connected programmes;
- Partnerships.

The UK is highly successful in securing EU research funding. The country contributes around 11% to the EU research budget and receives around 16% of the allocated funding (House of Lords 2016a, p. 7). Thus, the British R&D sector gets more from the EU R&D budget than it pays into it. From 2007 to 2013, the UK contributed EUR 5.4 billion to EU research projects, while receiving around EUR 8.8 billion back in the same period (Royal Society 2015, p. 12). Most of the EU funding in the UK comes from EU framework programmes. In the 2007–2013 period, 66% of the EUR 8.8 billion that the UK received from the EU research budget was financed through the 7th Framework Programme. Structural funds accounted for 22% and Marie Skłodowska-Curie Actions taken by UK researchers for mobility constituted 12% of total EU funding (Royal Society 2015, p. 13).

Taking together the EU grants under Framework Programmes and structural funds for research and innovation, the UK is in fourth place in the EU28 in terms of EU funds absorption, behind Germany, Poland and Italy (Fig. 2.3).

When it comes to EU framework programmes, the UK is the second largest recipient of research funding in the EU, after Germany. The UK is the first in the EU in terms of FP7 funds relative to GDP. In the 2007–2013 period, FP7 financing was equivalent to 3.2% of the GDP in the UK, 3.1% in Germany and 2.4% in France. Meanwhile, the UK was nowhere near the top of the list in terms of the per capita value of FP7 research grants it received from 2007 to 2013 on a competitive basis. Nine EU members—the Netherlands, Sweden, Denmark, Belgium, Finland, Austria, Ireland, Cyprus and Luxembourg—performed better than the UK in this respect (Table 2.4).

However, looking at the relative size of the economy, measured by the difference between the percentage of FP7 funding received by each member state and its proportion of EU GDP, it can be concluded that the UK performs quite well, being surpassed only by the Netherlands (Royal Society 2015, p. 15).

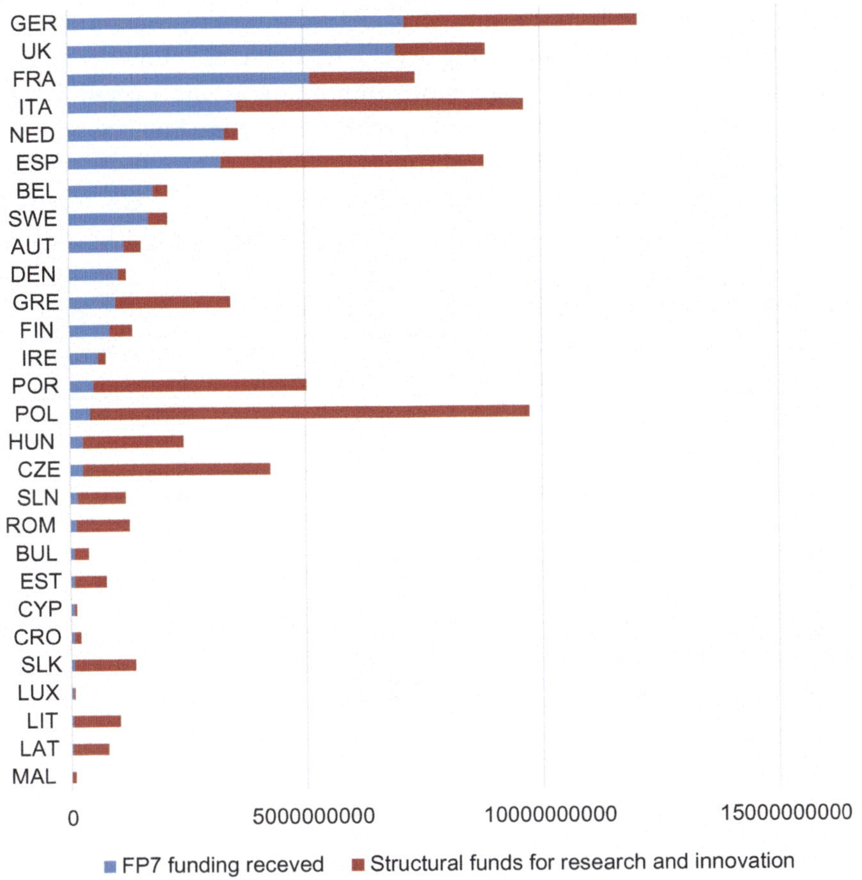

Fig. 2.3 Geographical distribution of EU expenditure on research, development and innovation (FP7 and structural funds) in EU28 countries, 2007–2013 (in EUR) (*Source* Own elaboration based on Royal Society [2015] and a database retrieved from https://publications.parliament.uk/pa/ld201516/ldselect/ldsctech/127/12702.htm. Accessed 18 October 2017)

The UK is also the leader in the EU when it comes to applications and participation in EU R&D programmes. As shown in Table 2.4, although Germany got the highest value of FP7 grants overall, the UK overtook Germany on Horizon 2020 with the highest share of signed grant agreements (European Commission 2015, p. 18; House of Lords

2016b, p. 7). These facts confirm that EU framework programmes are an important source of financing R&D in the UK. However, other sources of financing R&D are more important as the overall R&D budget that comes from EU research programmes constitutes about 3% of the country's total R&D expenditure (Fig. 2.4).

It should be pointed out, however, that businesses in the UK received around 18% of the FP7 funding in the UK, far below the EU average

Table 2.4 EU funds received under FP7 by EU28 countries from 2007 to 2013

Member state	FP7 funding received (in EUR million)	FP7 per capita (in EUR)	FP7 as % of GDP (in %)
Malta	20,930	48.8	0.02
Latvia	48,500	24.4	0.07
Lithuania	55,110	18.9	0.08
Luxembourg	60,390	107.3	0.01
Slovakia	78,420	14.5	0.09
Croatia	89,510	21.2	0.12
Cyprus	92,880	109.7	0.06
Estonia	94,260	71.8	0.11
Bulgaria	98,630	13.7	0.27
Romania	143,440	7.2	0.31
Slovenia	170,870	82.8	0.14
Czech Republic	288,450	27.4	0.28
Hungary	290,520	29.5	0.41
Poland	439,280	11.6	0.67
Portugal	520,700	50.2	0.45
Ireland	625,950	135.3	0.23
Finland	876,140	160.1	0.35
Greece	1,000,470	92.5	0.73
Denmark	1,060,600	187.4	0.35
Austria	1,184,210	137.9	0.47
Sweden	1,707,860	175.2	0.60
Belgium	1,814,890	161.2	0.77
Spain	3,256,330	70.1	2.01
Netherlands	3,329,970	197.0	1.25
Italy	3,589,820	59.0	1.90
France	5,142,710	77.5	2.36
UK	**6,940,060**	**107.2**	**3.22**
Germany	7,136,480	87.9	3.12

Source Own elaboration based on Royal Society (2015), https://royalsociety.org/~/media/policy/projects/eu-uk-funding/uk-membership-of-eu.pdf. Accessed 18 October 2017 and the database at https://publications.parliament.uk/pa/ld201516/ldselect/ldsctech/127/12702.htm. Accessed 18 October 2017

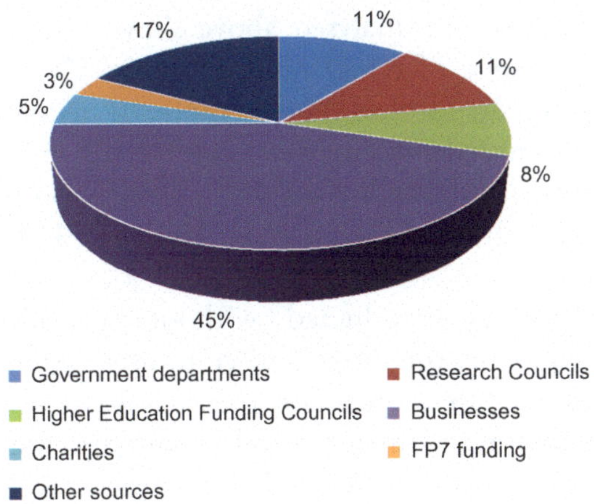

Fig. 2.4 The role of the EU's 7th Framework Programme (FP7) in UK expenditure on R&D, 2007–2013 (share of different sources of R&D funding, in %) (*Source* As in Table 2.4)

of 27%, and less than the UK's peers in Western Europe. In Germany, 33% of FP7 funding went to businesses, and in France, the figure was 27% (Royal Society 2015, p. 18). In the UK, 71% of FP7 funds were awarded to universities. This means that universities will probably be hardest hit if the UK is excluded from EU R&D funding after Brexit.

It is also worth looking at business R&D (BERD) in the UK from another perspective, i.e. by analysing the contribution of foreign-owned enterprises in the UK. The aim is to examine to what extent European-owned companies invest in R&D in the UK. Business R&D expenditure by foreign-owned businesses (i.e. enterprises with a single foreign owner holding over 50% of the shares) in the UK has grown in the last two decades. The same holds for expenditure by EU-owned businesses in the UK. Their share in total BERD in the UK has doubled during the last two decades, growing from 8% in 1995 to 15% in 2011. The role of non-EU businesses in total BERD in the UK increased even more. Their share, at 35% in 2011, was higher than that of EU-owned enterprises (Office for National Statistics 2017). It seems that non-European businesses active on the UK market have contributed more than EU-owned enterprises to the growing importance of foreign sources fuelling BERD in Britain.

The data presented and analysed above show that European funds are an important source of R&D funding in the UK, in particular in financing its university sector. Furthermore, a reduction in public research funding to universities between 2009/2010 and 2013/2014 did not decrease university research income. Just the opposite, this income increased over that period, largely due to growing EU funding (House of Lords 2016a, p. 12).

Research cannot be conducted without specialised infrastructure such as equipment, buildings and computer systems. Apart from national research facilities, there are pan-European research facilities funded by participating countries. The EU supports them through the European Strategic Forum on Research Infrastructures (ESFRI). Under FP7, the EU granted about EUR 300 million to all pan-European research facilities and further grants are being provided under Horizon 2020. Six pan-European research facilities are headquartered in the UK. These are:

- High Power Laser Energy Research Facility (HiPER)—Harwell, Oxfordshire (Central Laser Facility)
- ELIXIR (European Life-science Infrastructure for Biological Information)—Hinxton
- Integrated Structural Biology Infrastructure (INSTRUCT)—Oxford
- Infrastructure for Systems Biology-Europe (ISBE)—London (Imperial College)
- Square Kilometre Array (SKA)—Manchester (Jodrell Bank)
- European Social Survey (ESS ERIC)—London (City University) (Royal Society 2015, p. 20).

The UK also hosts 10 other pan-European research facilities that have headquarters in other European countries. FP7 projects that funded the involvement of EU members in pan-European research infrastructures saw the involvement of 146 different UK participants, with a total of 629 "participations," or 11.9% of the total number. The EU's contribution to these projects amounted to EUR 272.7 million, i.e. about 17.8% of the overall budget allocated to research infrastructures. Horizon 2020 projects related to research infrastructures launched by

2016 had included 78 participants from the UK on a budget of EUR 66.7 million, representing approximately 12.1% of the total budget allocated to these projects (European Commission 2016, pp. 8–9). These data confirm the relatively high involvement of the UK in pan-European infrastructures in terms of the number of participants and value of grants awarded through European framework programmes.

Summing up the analysis of the role of the EU in R&D funding in the UK, it can be concluded that EU funds have played a growing role in UK research. EU funding operates synergistically with domestic R&D funding sources. It leverages further domestic funding. The long EU funding cycle also provides stability for the research community. The UK is highly involved in EU funding programmes, having one of the highest rates of successful grant applications in the EU. European research programmes represent one of the largest sources of financing international networks globally and enable the establishment of large-scale, transnational and interdisciplinary collaborations. The implications of Brexit for R&D funding in the UK will depend on the rules of the new partnership between the UK and the EU in the post-Brexit period. Possible scenarios are presented in the next section of this chapter.

2.6 The Impact of Brexit on British R&D: Some Tentative Conclusions

According to an official statement by the British government on the UK's post-Brexit partnership with the EU, "it is the UK's ambition to build on its uniquely close relationship with the EU, so that collaboration on science and innovation is not only maintained, but strengthened" (HM Government 2017a, p. 7). However, the government also says that a strong partnership between the UK and the EU will be accompanied by a more outward-looking approach in order to achieve a truly global UK, i.e. "a country that reaches beyond the borders of Europe" (HM Government 2017b, p. 68).

Many studies on the potential impact of Brexit on the UK economy, including science and innovation, conclude that Brexit will bring some negative consequences for the UK and the rest of the EU, yet the costs and benefits will depend on the future model of collaboration (Ramphal 2016; Galbraith 2016; Jensen 2016; Minford and Pryce 2017). To adequately address the changing role of the EU in UK science, in particular R&D funding, at least two possible scenarios of UK–EU science collaboration can be presented and compared. Possible options for the UK's science relations with the EU are as follows:

1. Leaving the EU and the European Research Area (ERA), which will mean a "hard" Brexit for the science sector;
2. Acquiring Associated Country status in the EU science system in the post-Brexit period (similar to the model of the EU's collaboration in science with Switzerland or Norway).

Leaving the EU and the ERA means that the UK will have no access to Horizon 2020 and future EU framework programmes beyond third-country status. However, such a scenario could bring some additional money for UK science because the UK is a net contributor to the EU budget as a whole. The country's net contribution to the EU budget is equivalent to an estimated 0.6% of nominal GDP (House of Lords 2016a, p. 548). However, when it comes to the contribution to the European R&D budget, the analysis conducted in the previous section revealed that the UK gets more than it pays. As a result, in the short term, leaving the EU may lead to a reduction in R&D expenditures in the UK, but bring some surplus coming from the overall budget. Therefore, if the surplus released from the overall UK contribution to the EU budget is channelled into research and innovation, it can offset the loss of R&D funds that the UK receives through EU R&D programmes. However, that is unlikely to happen as all economic analyses model a loss in GDP for the UK in the transition years following Brexit (House of Lords 2016a, p. 549). The GDP decrease will probably cause a decline in public R&D expenditures in the UK. Therefore, UK research will suffer, at least in the short term, if a "hard" Brexit is implemented.

The second option, i.e. the UK acquiring full associated country status post-Brexit (similar to that enjoyed by Switzerland), will mean that the UK will leave the EU, but will still be included in EU science programmes. Such a scenario will require continued EU budget contributions. According to rough estimates, to maintain access to EU R&D programmes, the UK will have to pay more than its current contribution as the way of calculating it will probably change (it will be based on GDP and population size). Another option for the UK is not to pay more, but to accept limited involvement in EU research programmes. The UK will then have no right to shape the rules of EU financing and it will have to create additional domestic administration structures for managing EU programmes (House of Lords 2016a, p. 551). The complexity of different factors influencing the UK's involvement in EU programmes and additional costs related to them make it impossible to calculate the financial consequences of Brexit for UK research at this stage. However, as has been shown above, there will be some additional costs, while the benefits of the UK's participation in EU programmes will be somewhat limited. It is probable that the benefits that the UK science sector obtains will be reduced without the UK in the EU. As most of the EU R&D funds in the UK are awarded to universities (over 70%), research conducted by universities will probably be hardest hit by Brexit. Business R&D in the UK may suffer less because domestic enterprises are not strongly involved in EU programmes and because the foreign component of BERD in the UK benefits more from investment by non-EU enterprises operating in the British market rather than EU-owned businesses.

2.7 Conclusions

The UK has a strong track record in research and innovation, ranking fifth in the EU in terms of innovation performance measured by the Summary Innovation Index (SII). The UK's innovation performance relative to that of the EU has increased by 11.7% since 2010 (European Commission 2017b). As an EU member, the UK benefits from European research projects financed through Framework Programmes. Apart from claiming a high share of EU research funding, the UK has the largest number of participants in the Horizon 2020 programme.

This will probably change after Brexit. The impact of Brexit on research and innovation in the UK depends on the model of the future relationship between Britain and the EU. If the UK remains in the European Research Area, the consequences of Brexit for its STI sector may be limited. If, however, the UK is given third-country status, then the implications may be significant.

The most probable scenario for future research collaboration between the UK and the EU is that the UK will become a partially associated country. The EU has introduced and used the concept of partial association with Switzerland. Such a status means that the UK will contribute to EU framework programmes based on its GDP and population, but will have no role in developing EU research and innovation policy. In the long term, the UK may follow the same scenario as Switzerland, which experienced a decline in participation and financial benefits received from EU research programmes. In the UK, a key factor will be additional costs of re-negotiating access to EU research programmes after Brexit as well as costs related to involvement in EU programmes (e.g. the necessity to create new domestic administration structures). These may have a negative impact on the UK's research collaboration with EU partners post-Brexit.

References

Archibugi, D., & Iammarino, S. (2000). Innovation and Globalization: Evidence and Implications. In F. Chesnais, G. Ietto-Gilles, & R. Simonetti (Eds.), *European Integration and Global Corporate Strategies*. London and New York: Routledge.

Baldwin, R. E. (2015). *The Economics of European Integration*. London: McGraw-Hill Higher Education.

Barcz, J., Kawecka-Wyrzykowska, E., & Michałowska-Gorywoda, K. (2016). *Integracja europejska w okresie przemian. Aspekty ekonomiczne*. Warszawa: PWE.

Carlsson, B. (2006). Internationalization of Innovation Systems: A Survey of the Literature. *Research Policy, 35*(1), 56–67.

ERC. (2017). *ERC Starting Grants 2017: Outcome—Indicative Statistics*. https://erc.europa.eu/sites/default/files/document/file/erc_2017_stg_statistics.pdf. Accessed October 30, 2017.

European Commission. (2011). *Europe 2020 Flagship Initiative Innovation Union, SEC(2010) 1161*. Communication from the Commission to the European Parliament, the Council, the European Economic and Social Committee and the Committee of the Regions, Directorate-2011 General for Research and Innovation. Luxembourg: Publications Office of the European Union.

European Commission. (2015). *Horizon 2020: First Results*. Directorate-General for Research and Innovation. Luxembourg: Publications Office of the European Union. https://ec.europa.eu/programmes/horizon2020/sites/horizon2020/files/horizon_2020_first_results.pdf. Accessed October 30, 2017.

European Commission. (2016). *Written Evidence (EUM0081), European Commission Response to the Questions Addressed to Commissioner Moedas*. http://data.parliament.uk/writtenevidence/committeeevidence.svc/evidencedocument/science-and-technology-committee-lords/relationship-between-eu-membership-and-the-effectiveness-of-uk-science/written/30230.pdf. Accessed October 28, 2017.

European Commission. (2017a). *In-Depth Interim Evaluation of Horizon 2020, SWD(2017) 220 Final*. Brussels: Commission Staff Working Document.

European Commission. (2017b). *European Innovation Scoreboard 2017*. http://ec.europa.eu/growth/industry/innovation/facts-figures/scoreboards. Accessed September 20, 2017.

European Parliament. (2000). *Lisbon European Council 23 and 24 March 2000: Presidency Conclusions*. http://www.europarl.europa.eu/summits/lis1_en.htm.

Galbraith, J. (2016). Europe and the World After Brexit. *Globalizations, 14*(1), 164–167.

HM Government. (2017a). *Collaboration on Science and Innovation: A Future Partnership Paper*. London. www.gov.uk/government/publications. Accessed October 30, 2017.

HM Government. (2017b). *The United Kingdom's Exit from and New Partnership with the European Union*. www.gov.uk/government/publications. Accessed October 30, 2017.

House of Lords. (2016a). *Leaving the European Union: Funding for Universities and Scientific Research*. London: House of Lords Library. http://researchbriefings.parliament.uk/ResearchBriefing/Summary/LLN-2016-0055. Accessed October 20, 2017.

House of Lords. (2016b). *EU Membership and UK Science*. Oral and Written Evidence. London: Science and Technology Select Committee. https://publications.parliament.uk/pa/ld201516/ldselect/ldsctech/127/12702.htm. Accessed October 20, 2017.

Jensen, M. D. (2016). When Politics Prevails: The Political Economy of a Brexit. *Journal of European Public Policy, 23*(9), 1302–1310.

Long, T. B., & Blok, V. (2017). When the Going Gets Tough, the Tough Get Going: Towards a New—More Critical—Engagement with Responsible Research and Innovation in an Age of Trump, Brexit, and Wider Populism. *Journal of Responsible Innovation, 4*(1). https://doi.org/10.1080/23299460.2017.1319036.

Minford, P., & Pryce, V. (2017). The Prospects for Brexit: Two Views. *Economic Affairs, 37*(3), 422–427.

Niosi, J., & Bellon, B. (1994). The Global Interdependence of National Innovation Systems. *Technology in Society, 16*(2), 173–197.

Office for National Statistics. (2017). http://webarchive.nationalarchives.gov.uk/20160106012454/http://www.ons.gov.uk/ons/rel/rdit1/bus-ent-res-and-dev/2011/sty-berd.html. Accessed October 20, 2017.

Olbrich, P., & Witjes, N. (2014). Rethinking Borders of National Systems of Innovation: Austrian Perspectives on Korea's Internationalization of Green Technologies. *STI Policy Review, 5*(2), 65–95.

Ramphal, S. (2016). Brexit: 'Everybody Lost'. *The Round Table, 105*(5), 593–594.

Royal Society. (2015). *UK Research and the European Union: The Role of the EU in Funding UK Research.* https://royalsociety.org/~/media/policy/projects/eu-uk-funding/uk-membership-of-eu.pdf. Accessed October 18, 2017.

Schwang Serger, S., & Wise, E. (2010). *Internationalization of Research and Innovation—New Policy Developments.* Contributed Paper for the 2nd Conference on Corporate R&D (CONCORD—2010). http://iri.jrc.ec.europa.eu/documents/10180/12971/Internationalization%20of%20Research%20and%20Innovation%20%E2%80%93%20new%20policy%20developments. Accessed October 15, 2017.

UCEA. (2017). *Higher Education Workforce Survey 2017.* London: Universities and Colleges Employers Association.

UKCISA. (2017). https://www.hesa.ac.uk/data-and-analysis/publications/students-2015-16/introduction. Accessed September 30, 2017.

Weresa, M. A. (2014). *Polityka innowacyjna.* Warszawa: Wydawnictwo Naukowe PWN.

3

Brexit as a National Transformation Programme: Project Management Perspective

Ewa Sońta-Drączkowska

3.1 Introduction

It is commonly argued, in both the media and economic literature, that the UK and the European Union face a period of major change with Brexit, and that the outcomes are difficult to predict. Many ongoing government projects will be affected by this change. Brexit itself will trigger a range of additional projects necessary for the transformation, and many new projects will be carried out as a consequence of the EU's exit, impacting both the private and public sectors. The debate about the future of Britain and the EU after Brexit is ongoing and definitely not closed. There are still many questions about the future European landscape after Brexit. There is a lot of macroeconomic discussion in the literature about Brexit and its implications for the UK as well as European countries and the EU as a whole (Dhingra et al. 2016; Kierzenkowski et al. 2016; Goodwin and Heath 2016). However, little

E. Sońta-Drączkowska (✉)
Collegium of World Economy,
SGH Warsaw School of Economics, Warsaw, Poland
e-mail: esonta@sgh.waw.pl

A. M. Kowalski (ed.), *Brexit and the Consequences for International Competitiveness*, https://doi.org/10.1007/978-3-030-03245-6_3

attention has been paid to how the Brexit process will actually be executed. Since this is a large and complex undertaking, the way in which Brexit will be carried out may have a major influence on the outcomes of the project, its success or failure. This chapter discusses Brexit from the project management perspective.

3.2 Study Goal and Methodology

The purpose of this study is to analyse Brexit as a national transformation programme from the perspective of project management discipline. The key aspects of the transformation are analysed as well as the role of project management in this process. Several questions and issues are raised based on programme management methodologies as well as literature on managing complex projects in the public sector. Since the author did not find any papers on this topic in academic journals, the study is based on a review of existing project management frameworks and project management literature related to managing public sector projects as well as secondary resources and internet research on Brexit. Special attention is given to two areas: governance issues and civil service resources. The Managing Successful Programmes (MSP) guidance, which is a compendium of programme management best practices (MSP 2011), is the analytical framework for the study. Public value management theory (PVM) will be a theoretical perspective for the analysis (Considine and Lewis 2003; Stoker 2006). The research question is: What are the challenges of Brexit as a national transformation programme from the project management perspective?

3.3 Brexit as a Transformation Programme

A vast majority of EU citizens were surprised by the outcome of the Brexit referendum on 23 June 2016, when 51.9% of Britons voted in favour of their country leaving the EU. There was no plan B for the situation that the British people would vote "Leave" (Evans 2018). Thus the outcome of the vote surprised Britain and its government, leaving the country facing an uncertain future. On 29 March 2017, the British government invoked Article 50 of the Treaty on the European Union; although revoking this might be legally possible, the UK is thus on course to leave the EU by March 2019. Brexit is probably the biggest short-term change management programme in the UK since World War II. The government has a huge pipeline of major projects planned or already under way, boiling down into a large nationwide transformation programme.

Project management discipline as a sub-segment of the organisational theory is associated with management of change. Projects in the organisation are ways to implement strategy through a coordinated set of changes. Changes in a political environment are no different. We can assume that implementing change in a democracy on a national and international scale is far more complex than implementing change within an organisation. Thus executing Brexit will be a transformation challenge. From the project management perspective, Brexit is not a single project but a group of interrelated projects. Therefore it cannot be perceived as a single undertaking but as a programme. The programme is defined as "*a temporary flexible organization structure created to coordinate, direct, oversee the implementation of a set of related projects and activities in order to deliver outcomes and benefits related to organisation's strategic objectives*". Meanwhile, programme management is "*the coordinated organization, direction, and implementation of a dossier of projects and transformation activities (i.e. the programme) to achieve outcomes and realize benefits of strategic importance*" (MSP 2011, p. 4).

The Brexit programme is characterised by a relatively low predictability of results and the scope of changes will impact entire society and its environment (MSP 2011, p. 8). With such programmes, it is highly probable that scope alignment will be necessary during the implementation stage as more details unfold. Therefore the Brexit programme has an emergent character. It has all the features of a large public sector transformation. According to MSP guidelines, such a programme may last for years and needs adjustments after each tranche is completed. In this chapter, two aspects of the Brexit programme are discussed: (1) programme governance issues and (2) Brexit programme resources.

3.4 The Brexit Transformation Environment: A Theoretical Perspective

To understand the environment and social context of the Brexit programme, a brief historical perspective needs to be outlined of how the public service model evolved. Until the 1980s the traditional public service model was the dominant paradigm in the civil service, with a focus on procedural governance, planning and top-down policy implementation. The next two decades were dominated by the so-called New Public Management (NPM) approach, focusing on corporate and, to a lesser extent, market governance of the public service. A nexus of contracts was designed to ensure efficiency for the public sector. From early 2000 onwards, there were shifts from the corporate orientation towards a more "network oriented" model, referred to also as Public Value Management (Moore 1995; Meynhardt 2009). Network governance is "*a particular framing of collective decision making that is characterized by a trend for a wider range of participants to be seen as legitimate members of the decision-making process in the context of considerable uncertainty and complexity*" (Stoker 2006, p. 41). In the network governance model, we can expect the trend towards public participation and consultation, political responsiveness and flexibility (Crawford and Helm 2009). In this context, Brexit can be perceived as a manifestation of this theory in practice, where citizens decide about the future of the nation. According to Meynhardt (2009): "*Public value is value for the public. Value for the public is a result of evaluations about how basic needs of individuals, groups and the society as a whole are influenced in relationships involving the public. Public value then is also value from the public, i.e., 'drawn' from the experience of the public*".

Moore (1995) defines public value in the following way: "*The definition that remains equates managerial success in the public sector with initiating and reshaping public sector enterprises in ways that increase their value to the public in both the short and the long run*".

Network governance and orientation on "public value" pose additional challenges to Brexit implementation. In line with this theory,

Table 3.1 Development of the public governance model: setting the scene for Brexit transformation

	Traditional Public Management (until 1980s)	New Public Management (NPM) (early 1980s–2000s)	Public Value Management (PVM) (from 2000)
Model of Governance	Procedural	Corporate	Network
Mode of Operation	Planning and policy	Management and contracts	Knowledge fields
Performance Objective	Managing inputs	Managing inputs and outputs	Multiple objectives: Service outputs; satisfaction; outcomes; trust and legitimacy
Goal of Managers	Respond to political direction	Meet agreed performance targets	Respond to citizen preferences, renew mandate and trust through quality services
Accountability	Upwards through departments to politicians to parliament	Upwards through performance contracts; sometimes outward to customers through market mechanisms	Multiple: Citizens as overseers of government; customers as users of services; taxpayers
Role of Community	Little community involvement	Increased consultation	Community enablement and involvement

Source Based on Crawford, L., & Helm, J. (2009). Government and Governance: The Value of Project Management in the Public Sector. *Project Management Journal, 40*(1), 73–87

Brexit leaders should act as wise entrepreneurs and ensure that Brexit contributes to short and long-term public value creation. Public value management orientation takes into account multiple objectives—such as creating trust and legitimacy, responding to citizens' preferences, community involvement and multiple accountability—and allows citizens to oversee government. How, in the light of the theory, will the government be able to satisfy British citizens voting "Leave" and those voting "Stay" at the same time? What should be the orientation point of the "public value"? In this challenging environment, aligning the interests of a divided British nation will probably be a tough task, and a range of compromises as well as "lose-lose" deals can be expected (Table 3.1).

3.5 Programme Governance Issues

The term "governance" in the organisational context is essentially associated with accountability and responsibilities. It is defined here as the functions, processes, procedures and responsibilities that provide the framework for how the programme is to be set up, managed and controlled. Programme governance provides the backdrop for all activities of managing the programme and achieving the programme's outcomes (MSP 2011).

According to the "Managing Successful Programmes" (MSP) methodology, there are nine programme governance aspects that should be taken into account while setting up and managing a programme: (1) programme organisation, (2) vision, (3) leadership and stakeholder engagement, (4) benefits realisation management, (5) blueprint design and delivery, (6) planning and control, (7) business case, (8) risk and issue management, (9) quality management (MSP 2011). Based on the MSP framework, the most important themes are discussed in the context of Brexit: first, vision, the business case behind the transformation and transformation benefits; second, programme leadership and stakeholder engagement; and third, programme blueprint design and implementation.

3.6 Vision, Business Case and Transformation Benefits

The Brexit programme lacks a clear vision. The MSP guidance specifies that a programme should have a clear "Programme Vision Statement". According to that document programme, the vision needs to have several characteristics: it should be clear, attractive, beneficial and motivating; it should show a better future; be easy to communicate; and finally short and easy to memorise. In the case of Brexit, it is difficult to say that the vision has the characteristics of a good programme vision statement. Several options are discussed as possible visions for the UK after Brexit (Busch and Matthes 2016). British Prime Minister Theresa May has said during debates that neither the Canadian nor the Swiss model would be appropriate for managing Britain's future relations with the EU, but she has encouraged being "*creative and practical*" in drawing up a completely new set of arrangements for British–EU relations in the future (September, 2017). Thus, it may be concluded that a year-and-a-half before "day one", the vision was blurred, uncertain or even non-existent.

The Brexit programme lacks a compelling rationale in terms of both clear benefits and the business case. The business case for the programme as well as the definition of its key benefits is the foundation for starting a programme. A rationale for such an undertaking in the business context is a must (MSP 2011). However, in the public sector, the programme is usually set up to solve a complex social problem, capture an opportunity or provide development support for a particular region or social group. From this perspective, public programmes do not necessarily have to be "profitable" in a business sense, but they have to aim at achieving "public value". In the context of the Public Value Management theory, there is one important question that does not have a simple answer: Who is the beneficiary of the Brexit programme? How should the "public value" from Brexit be measured? Initial macroeconomic analysis shows that, due to Brexit, the average UK incomes will fall by anywhere from 1.3% (or GBP 850 per household) in a best-case scenario to 2.6% (GBP 1700 per household) in a worst-case scenario.

One of the expected costs of Brexit is that the UK stands to lose anywhere from 6.3% to 9.5% of GDP, or about GBP 4200 to GBP 6400 per household (Dingra et al. 2016). It is still uncertain what aggregate influence Brexit will have on Britain. Based on preliminary macroeconomic studies, losses are more probable than gains.

In the case of Brexit, the rationale for the move was a combination of emotional and psychological rather than rational factors. It was correlated with the emergence of populism in Europe and a backlash against cultural change in Western Europe in recent decades (Goodwin and Heath 2016). Britons voting "Leave" were rebelling against bureaucracy in Brussels, taking a stand on the refugee issue, guarding traditional values and fearing terrorism. Economic calculation was not a key argument in the battle. From this perspective, there are still several open questions. One question is: What are the key public values (benefits) of Brexit for "Leave' voters?" Another question is: How will British negotiators take this into account when carrying out the Brexit programme? If they are going to approach the needs of Leave voters based on emotions and fears, Britain will probably follow a path of real separation from core European principles. This is one of the key questions the British programme task force will have to face during Brexit negotiations.

3.7 Programme Leadership and Stakeholder Engagement

Brexit lacks a strong and charismatic leadership (programme sponsor). One of the most important characteristics of a successful transformation programme is having an energetic, engaged and charismatic sponsor who strongly believes in the idea. According to the MSP, programme leadership is a key requirement because it helps manage change, achieve steps necessary for the transition and influence people on the way towards the future vision. That is why a strong leader needs to be more than an administrator and manager. He should have a strong vision and be credible in his position. In addition, strong negotiation skills will be priceless for Brexit. German Chancellor Angela Merkel has made it clear that the European Union will everything possible to ensure that Britain's separation from the EU will not be a "cherry-picking exercise".

More than a year before Brexit day, it could be expected that Britain's position at the negotiating table would not be particularly strong. The two people who might have been credible as Brexit leaders—Boris Johnson and Nigel Farage, strong promoters of the UK leaving the EU—resigned from leadership positions shortly after the referendum vote. The question was whether Theresa May had enough experience to lead large transformations. And it was widely known that she was more inclined towards the "remain" option during the referendum. Many doubted if she would prove to be credible as a leader of Brexit and whether she would be able to put together a strong leadership group to carry out tough negotiations.

In April 2017, May announced an early general election with the aim of strengthening her hand in the Brexit negotiations. The June 2017 election resulted in a hung parliament in which the number of Conservative seats fell from 330 to 317, prompting May to broker a confidence and supply deal with the Northern Ireland's Democratic Unionist Party (DUP) to support her minority government.

Another issue related to programme leadership is the necessity of winning stakeholder engagement and support for the programme. The most successful programmes occur when all, or at least most of the key stakeholders (people, organisations, institutions impacted by the programme, who have some level of influence and/or interest in the programme), are in favour of the change. With Brexit, half the country appears to be against the change, and many will do all they can to make the programme fail. Thus, it will be probably a challenging exercise to identify, plan and manage the nexus of stakeholders affected by Brexit. To mention only some of them:

- Brexit task force running negotiations both on the EU side in Brussels and in the UK,
- European institutions: Council, Commission, Parliament (and relevant regulatory bodies) and the British government (and relevant regulatory bodies),
- European Union and 27 EU countries,

- British citizens (divided into two major subgroups: those voting "Leave" and those voting "Remain"),
- Immigrants leaving and working in the UK,
- Scotland and Wales, which are affected, but might have different interests,
- Other Commonwealth countries,
- Citizens of 27 EU countries,
- Governments of 27 EU countries,
- Business and private companies in the UK and Europe,
- 168 other, non-EU countries which will be affected by the UK leaving Europe.

The list goes on. One thing is clear, that the Brexit programme characterises the huge complexity and involvement of many groups, with different, often conflicting, interests and agendas. The environment is highly political and affects not only Britain but all of Europe and the network of international economic and political relations. According to the MSP methodology, programme management stakeholders need to be identified, analysed and mapped according to two dimensions: (1) interest in programme outcomes and (2) influence on the programme. This initial assessment makes it possible to identify key stakeholders and draw up appropriate strategies to ensure that the interests and expectations of stakeholders are taken into account. Special attention should be paid to programme opponents, who may be an impediment to programme success. A key issue in managing stakeholders is communication on programme goals, benefits and work in progress (MSP 2011, pp. 54–64). From the perspective of public value management theory (PVM), the priority for the Brexit task force should be the "British citizen" because the government serves the nation. The government needs to meet multiple objectives and listen to the voice of the community and allow communities to be "legitimate members of the decision-making process". Which section of the British public should the government serve better? How should it balance the interests of those who are for and those who are against Britain leaving the EU? Stakeholder engagement and communication will be a key issue for Brexit to succeed. Additionally, Britain's Brexit task force should be prepared for a tough battle with EU officials because the EU needs to hold a common front in order to discourage other member countries against taking similar action.

3.8 Programme Blueprint Design and Implementation Time Frame

Brexit is missing a programme blueprint. In the project management methodology, the scope defines the product of the project and all the necessary tasks to deliver it. The more we know about this at the beginning of the project, the better we can estimate the duration, cost and skills needed to produce the desired outcome. Conversely, the more uncertainty there is about the scope, the more difficult it is to have an accurate plan. The Brexit programme is unique, and defining the scope of this undertaking will be a painful task that will take months and maybe years. The longer it takes, the higher the chances of failure. The MSP describes the programme scope in terms of a "blueprint". A blueprint is a detailed vision for an organisation (or to-be situation), covering what the organisation will look like when all the projects are completed, the programme is wound up, and the business transformation is done. An approach supporting programme scope definition is the POTI model, which stands for processes, organisation, technology and information. These four areas make up a comprehensive view of all the elements that form the programme scope (MSP 2011). How can a detailed Brexit programme blueprint be prepared if we do not even know the vision? What concrete tasks are assigned to whom to make work progress manageable?

Some very general scope-related data are available. We know that Brexit will be conducted in three phases (tranches), which are described in high-level statements, without any precise milestones, work packages and tasks necessary to complete each phase. We know that after Brexit Britain will have to negotiate at least 759 treaties, including 295 on international trade, 202 on regulatory cooperation, 69 fisheries and 65 on transportation issues (FT 2017) (Table 3.2).

Table 3.2 Three Phases of Brexit transformation

1. Preparing for negotiations	2. Managing the negotiations	3. Implementing Brexit
• Coordinating analysis across Whitehall • Capturing the views of businesses, the devolved administrations and other interests	• With the EU institutions and 27 member states • On the terms of the UK's departure from the EU • On the new framework for relationships	• Transferring the existing *acquis* into UK law • Designing and implementing new domestic policies or regulatory regimes to replace EU functions • Preparing and pursuing a new international trade policy

Source The Civil Service after Article 50, Institute for Government, March 2017, p. 3

The Brexit programme lacks the precise setting of beginning and finish dates. Deadlines in project management are used to focus the team, increase their attention towards the end goal and, of course, to put pressure on them to perform at their best. Starting a project without a clear finish line can make a project drag on for months or even years. The symbolic deadline for Brexit is 29 March 2019 (referred to as "day one"). But it is clear that the programme will not finish by this date. It may even serve as the beginning date. There are already visible signals that some crucial projects that need to be completed before day one are troubled. One example is new regulations on customs agreements. Customs is undoubtedly one of the challenges, if not the biggest one, facing the government in avoiding the cliff edge scenario and implementing Brexit. Customs is an exclusive EU competence, which means that the treatment of goods at the UK border has been determined in Brussels in recent decades. Key challenges in this area include policy and legislation issues as well as preparation and implementation of necessary technology and customs infrastructure on time. According to a group of analysts, "*successful change relies on all these organisations being ready. In the past they have been given years to adapt to any government change; they now have fewer than 20 months to prepare without yet being clear what they are preparing for*" (Owen et al. 2017).

3.9 Civil Service Resources and Brexit

The importance of project and programme management capability in the public sector has been recognised as a key success factor in government initiatives in various countries (Considine and Lewis 2003). Brexit will create a multi-project environment in the British public sector. The challenges of managing a multi-project environment are broadly discussed in research reports (e.g. Zika-Viktorsson et al. 2006; Engwall and Jerbrant 2003; Elonen and Artto 2003). Researchers have examined a wide range of side effects of a multi-project environment, most of them related to the "human side" of project management, e.g. a lack of resources, competencies and methods, a lack of team commitment, unclear roles and responsibilities and inadequate information management (Elonen and Artto 2003). Some researchers have even introduced the notions of "projectification" and "programmification" of organisations and societies and debated their negative effects on individuals as well as on the efficiency and effectiveness of projects implementation in organisations (Maylor et al. 2006; Packendorff and Lindgren 2014).

In the light of these developments, Brexit can be perceived as an additional driver for a wave of projects initiated in the public sector. Too many project initiatives may cause disorientation, a lack of priorities, a resource overload, and as a result project implementation bottlenecks. Several analytical reports discussing the challenges of Brexit from the perspective of resource availability have been issued by the UK National Audit Office (NAO, March 2017). According to this institution, problem areas include the "capacity" and availability of staff as well as "skills", understood as ensuring that civil servants in the UK have the right skills to do their jobs properly. A National Audit Office report on the capability of the civil service (NAO, March 2017) found that "*leaving the EU will further increase capability challenges facing government.*" The report concludes that "*the decision to leave the EU also means government will take on tasks previously performed by others, requiring the development of skills not previously planned for*". The overall audit score of the civil service was 2.1 out of a maximum of 5. Below there is a selection of key findings from the civil service audit report:

- Civil servants are responsible for an increasingly complex range of tasks and projects. The government is asking to deliver more while the number of civil servants was reduced by 26% from 2006 to 2017. With the emerging requirements of Brexit, a significant resource gap can be expected.
- Weaknesses in capability undermine the government's ability to achieve its objectives. The audit report shows that many delivery problems can be traced to weaknesses in capability. The collapse of the InterCity West Coast franchise competition in 2012 was provided here as an example of a historical project that failed due to missing capabilities in project and programme management in the UK's civil service.
- Government projects too often go ahead without the government knowing whether departments have the skills to deliver them. While the civil service has skilled people, many of these projects draw on the same pool of skills.
- Departments do not know what skills they have, whether these are in the right place, and what additional skills they need. This means the government does not know enough about who is doing what and when, and whether those carrying out tasks have the necessary skills.

Leaving the EU will further increase the capability challenges facing the government. The Cabinet Secretary has referred to the UK's decision to withdraw from the EU as "*the biggest, most complex challenge facing the civil service in our peacetime history*". As a response to this challenge, the government has staffed up two new departments to support this process. The capability demands go beyond these two departments, and the civil service is currently evaluating the longer-term impacts of Brexit on its staffing needs. For example, departments that have had large amounts of EU-derived funding and legislation—such as the Department for Environment, Food and Rural Affairs—need legal, economic and sector experts to deal with the implications of Brexit. They will have to do this while using their remaining staff to achieve pre-existing priorities (NAO, March 2017).

The audit report concludes that the government has based its plan on growing skills in the civil service, and these will take time to develop.

The scale of the challenge means there is a need for greater urgency (NAO, March 2017). The following bundles of competencies are necessary in the British civil service:

- Analysis: civil servants will need to develop options for new policies, advise ministers and react to EU negotiating positions.
- Coordination: civil servants must be able to access the expertise of a range of groups, including devolved administrations, local government and businesses.
- Legislation: the civil service will need the skills and capacity to prepare, draft and manage the passage into law of a big body of legislation, some of which will be highly contentious and to tight timelines.
- Delivery: civil servants will then have to implement the outcome of any final deal, which includes new regulatory regimes, immigration systems and customs checks at UK borders (Table 3.3).

Several recommendations towards improving civil service readiness for implementing the Brexit programme were suggested:

- Recommendation 1: The government should provide departments with more detailed information on the timelines to which they are working, and on what will be required from departments by the time Article 50 is triggered.
- Recommendation 2: The government should decide as soon as possible how negotiations will be run and who will be involved.
- Recommendation 3: The government should ensure all departments are doing sufficient post-Brexit planning, including how to realise the opportunities offered by Brexit.
- Recommendation 4: The government should set out its priorities, ensure there is enough staffing and money for Brexit and existing commitments, or acknowledge where plans may be trimmed (Owen and Munro 2016).

Table 3.3 Brexit capacity demands

Key demands	Preparing for negotiations	Managing negotiations	Implementing Brexit
Analysis and policymaking	• Understand the strategic interests of the UK across sectors and feasibility of different options	• Capacity within departments and key public bodies to analyse and make recommendations to the negotiating team on emerging deals	• Develop new policy options in areas such as fisheries, agriculture, immigration and trade
Coordination	• Understand the views and preferences of different departments and public bodies, as well as local and devolved government and business	• Bring together insight from departmental policy experts on negotiating options, and using UKRep and Foreign Office intelligence on the views of EU member states	• Coordinate post-Brexit policy changes with devolved administrations. Diplomatic capacity to maintain links with EU institutions and new trade partners
Legislative capacity	• Departments must identify which EU rules and regulations should be transposed into UK law	• Ensure the "Great Repeal Bill" is prepared, drafted, scrutinised and approved to avoid a legal 'cliff edge' after Brexit	• Prepare and scrutinise potentially hundreds of pieces of primary and secondary legislation to enact post-Brexit policies
Delivery and implementation	• Major policy and reform commitments may need to be delayed to free up capacity for Brexit work	• Prepare new policies and administrative systems for day one after leaving the EU	• Delivery and implementation of post-Brexit policies including negotiating trade deals and procuring new IT services

Source The Civil Service After Article 50, Institute for Government, March 2017, p. 10

3.10 Conclusions

This chapter has examined Brexit from the perspective of project management discipline. Brexit is a complex undertaking, and it will trigger a range of projects as a result of the UK leaving EU, which will affect both the public and private sectors. The projects are related to the common goal and mutually connected, which means that Brexit is a "complex transformation programme". The Public Value Theory and Managing Successful Programmes body of knowledge provided the analytical framework for the study. Several aspects of Brexit as a national transformation programme were discussed, which can be divided into two major areas: (1) governance issues and (2) resource capacities of the British civil service. The main programme governance challenges are a lack of programme vision, lack of business case and transformation benefits, weak programme leadership and stakeholder engagement as well as a missing Brexit programme blueprint design and implementation time frame. Some major Brexit resource-related issues are the availability of staff (capacity) as well as competencies (project and programme management skills) of the civil service needed to carry out the transformation. The analysis led to the conclusion that, from the project management perspective, Brexit can be perceived as a high-risk programme with several shortcomings concerning proper programme set-up and management. The risk increases due to the limited capabilities of the British civil service. Facing this national transformation will require a range of both soft and hard skills from British civil servants combined with strong leadership and negotiation skills on the government level. Many question marks make programme preparation difficult and challenging. Is the Brexit programme a "mission impossible" or will it lead to success? It can be expected that Brexit *post-mortem* analysis will not be easy considering the lack of *ex-ante* success criteria and the complex setting of Brexit stakeholders.

References

After Brexit: The UK Will Need to Renegotiate at Least 759 Treaties. *Financial Times* (2017). Financial Times Research. https://www.ft.com/content/f1435a8e-372b-11e7-bce4-9023f8c0fd2e?mhq5j=e6.

Busch, B., & Matthes, J. (2016). *Brexit—The Economic Impact: A Meta-Analysis* (IW-Report No. 10/2016).

Capability in the Civil Service. (2017). *NAO National Service Office* (HC 919 SESSION 2016–17 24 MARCH 2017). https://www.nao.org.uk/wp-content/uploads/2017/03/Capability-in-the-civil-service.pdf.

Considine, M., & Lewis, J. M. (2003). Bureaucracy, Network, or Enterprise? Comparing Models of Governance in Australia, Britain, the Netherlands, and New Zealand. *Public Administration Review, 63*(2), 131–140.

Crawford, L., & Helm, J. (2009). Government and Governance: The Value of Project Management in the Public Sector. *Project Management Journal, 40*(1), 73–87.

Dhingra, S., Ottaviano, G., Sampson, T., & Van Reen, J. (2016). *The Consequences of Brexit for UK Trade and Living Standards*. London School of Economics.

Elonen, S., & Artto, K. A. (2003). Problems in Managing Internal Development Projects in Multi-project Environments. *International Journal of Project Management, 21*(6), 395–402.

Engwall, M., & Jerbrant, A. (2003). The Resource Allocation Syndrome: The Prime Challenge of Multi-project Management? *International Journal of Project Management, 21*(6), 403–409.

Evans, A. (2018). Planning for Brexit: The Case of the 1975 Referendum. *The Political Quarterly, 89*(1), 127–133. https://www.instituteforgovernment.org.uk/sites/default/files/publications/IFGJ5003_Whitehalls_preparation_131216_V10.pdf.

Goodwin, M. J., & Heath, O. (2016). The 2016 Referendum, Brexit and the Left Behind: An Aggregate-Level Analysis of the Result. *The Political Quarterly, 87*(3), 323–332.

Kierzenkowski R., Pain, N., Rusticelli, E., & Zwart, S. (2016). *The Economic Consequences of Brexit: A Taxing Decision*. OECD.

Managing Successful Programmes. (2011). London, UK: The Stationery Office, Axelos.

Maylor, H., Brady, T., Cooke-Davies, T., & Hodgson, D. (2006). From Projectification to Programmification. *International Journal of Project Management, 24*(8), 663–674.

Meynhardt, T. (2009). Public Value Inside: What Is Public Value Creation? *International Journal of Public Administration, 32*(3–4), 192–219.

Moore, M. (1995). *Creating Public Value—Strategic Management in Government*. Cambridge: Harvard University Press.

Owen, J., & Munro. R. (2016). *Whitehall's Preparation for the UK's Exit from the EU*. Institute for Government.

Owen, J., Shephard, M., & Sojanowic, A. (2017). *Implementing Brexit: Customs*. IGF Analysis, Institute for Government Commerce.

Packendorff, J., & Lindgren, M. (2014). Projectification and Its Consequences: Narrow and Broad Conceptualizations. *South African Journal of Economic and Management Sciences, 17*(1), 7–21.

Stoker, G. (2006). Public Value Management: A New Narrative for Networked Governance? *The American Review of Public Administration, 36*(1), 41–57.

The Civil Service After Article 50. (2017). Institute for Government. https://www.instituteforgovernment.org.uk/sites/default/files/publications/IFGJ5327_Report_Brexit_Civil_Service_080317_WEB.pdf.

Zika-Viktorsson, A., Sundström, P., & Engwall, M. (2006). Project Overload: An Exploratory Study of Work and Management in Multi-project Settings. *International Journal of Project Management, 24*(5), 385–394.

4

The Economic Order of Post-Brexit Europe and the Role of Germany: An Ordoliberal Perspective

Jürgen Wandel

4.1 Introduction

The completely unexpected decision by the British public to vote for Brexit in the 23 June 2016 referendum raised the question of the future shape of the European Union. After an initial shock, the European Commission in March 2017 reacted and issued a White Paper on the future of Europe in which it presented five scenarios for discussion on how Europe could evolve by 2025. Implicitly, the White Paper revealed a preference for a great leap forward towards a political union (Berthold 2017; Eder et al. 2017), which was subsequently underlined by European Commission President Jean-Claude Juncker in his State of the Union Address on 13 September 2017.[1] The ideas of a separate

[1]Accessible at http://europa.eu/rapid/press-release_SPEECH-17-3165_en.htm.

J. Wandel (✉)
Collegium of World Economy,
SGH Warsaw School of Economics, Warsaw, Poland
e-mail: jwande@sgh.waw.pl

A. M. Kowalski (ed.), *Brexit and the Consequences for International Competitiveness*, https://doi.org/10.1007/978-3-030-03245-6_4

budget for the eurozone, a common finance minister and a joint parliament, put forward by French President Emmanuel Macron, also point towards a "more Europe" trajectory.

The aim of this chapter is to analyse these options of the White Paper process for the future character of the EU's economic order and the role of Germany in it. An ordoliberal perspective is applied for this purpose. This is interesting for two reasons. First, already in the 1950s ordoliberal economists and politicians in Germany, in particular Wilhelm Röpke and Ludwig Erhard, debated about the economic and political constitution of Europe. Many of their arguments are still strikingly topical today in times of Brexit. Second, Germany is the biggest and economically most powerful EU member country. It was a key player in the eurozone crisis, and its weight will increase in an EU without the UK. Thus, Germany is expected to play a leading role in shaping post-Brexit Europe (see, e.g., *New York Times*, July 4 2016). This raises the question of whether Germany itself favours a distinct model of European integration after Brexit. Some scholars contend that Germany might advocate a governance structure with a strong ordoliberal flavour (Blyth 2013, p. 142; Biebricher 2014; Cardwell and Snaith 2018). Ordoliberalism provided the theoretical foundation of Germany's post-World War II economic system, known as a social market economy.

In order to assess the European Commission's post-Brexit EU scenarios as well as Germany's real-world economic policy preferences at the national and supranational levels, this chapter develops an ordoliberal reference model of European integration as an analytical framework. It will be argued that the ordoliberal reference model is largely supportive of the British vision of a free market-oriented and decentralised economic order of the EU rather than of an ever closer union as advocated by France and the European Commission. Nevertheless, Germany is unlikely to promote an inherent ordoliberal economic order in the EU for two reasons: first, Germany's own economic policy has increasingly departed from the ordoliberal ideal. Second, the "sacralisation" of the European project in German politics and public opinion makes it almost taboo to question the trajectory of European integration towards a federation, even if criticism was due from an ordoliberal point of view.

The remainder of this chapter is organised as follows. After an outline of the underlying understanding of ordoliberalism (Sect. 4.2), Sect. 4.3 develops the ordoliberal reference model. Through this lens, Sect. 4.4 analyses four post-Brexit EU scenarios, and Sect. 4.5 explores the likelihood that Germany might advocate reshaping the EU post-Brexit using the ordoliberal reference model. The chapter ends with concluding remarks (Sect. 4.6).

4.2 Definition of Ordoliberalism

Ordoliberalism is a branch of classical liberalism that evolved during the Nazi period in the 1930s around economist Walter Eucken (1891–1950) and two jurists, Franz Böhm (1895–1977) and Hans Großmann-Doerth (1894–1944), at the University of Freiburg in Germany. It is therefore also known as the Freiburg School, although other prominent ordoliberal thinkers worked at other universities, such as Wilhelm Röpke in Marburg and Geneva and Alexander Rüstow in Heidelberg.

The ultimate aim of their research programme was to seek an economic system that would provide for a high degree of human dignity and prosperity given the negative experiences with the planned economies of Nazi Germany and the Soviet Union, interventionism during the Weimar Germany period and in the period before World War I, which they labelled laissez-faire, when the formation of cartels to restrict competition received support from policy-makers and the courts. The research concept of ordoliberalism rests on the premise that the institutional framework—or "the order" (*ordo* in Latin)—is crucial for the nature of the economic and societal system as it determines the range of individual freedom of action and structures incentives underlying individual action. According to ordoliberals, individual freedom and prosperity can best be ensured in what is called a competitive order (*Wettbewerbsordnung*). The task to set up and ensure an appropriate institutional framework for it is attributed to a strong but limited government.

This focus on the institutional framework of economic life as well as the importance of competition is a point of interest that ordoliberalism has in common with classical liberalism,[2] the Austrian School as well as New Institutional Economics and Public Choice Theory. A growing body of empirical literature on why some societies are rich and others poor underscores the ordoliberal insight into the central importance of the right institutional set-up for prosperity (e.g. North 1990; Olson 2000; Acemoglu and Robinson 2012). Hence, it is difficult to clearly delineate ordoliberalism from these related research programmes and denounce it as a thing of the past, an ideology or a German *Sonderweg* (special path). Rather, much of the early Freiburg School can be regarded as preceding today's New Institutional Economics and Public Choice Theory (Streit and Wohlgemuth 1997; Goldschmidt et al. 2009; Wohlgemuth 2013a, b). Meanwhile, a younger generation of German economists (e.g. Manfred Streit, Viktor Vanberg, Michael Wohlgemuth, Lars Feld, Nils Goldschmidt, Gerhard Wegner and Joachim Zweynert) advances the ordoliberal approach through the incorporation of Hayekian Austrian economics,[3] the New Institutional Economics of Douglass C. North and Elinor C. Oström, and the Public Choice Theory of James M. Buchanan. This modernised, broader understanding of ordoliberalism underlies the ordoliberal reference model for European integration and the subsequent discussion of the positions of the European Commission and Germany on the EU's post-Brexit economic order.

---

[2]Classical liberalism, also known as 'old liberalism' or 'liberalism in the European sense', is not the same as North American liberalism ('new liberalism' or 'modern liberalism'). Although both share a strong belief in personal freedom, American liberalism gives much paternalist and interventionist power to the state similar to the European Social Democrats (Butler 2015).

[3]In 1962, this most prominent representative of the Austrian School took the chair for economic policy at Freiburg University and his insights are also sometimes referred to as the *Hayekian Ordnungstheorie* (Wohlgemuth 2001, p. 214; Goldschmidt 2006, p. 17). In particular, his view of competition as a discovery procedure has become widely accepted by younger economists in the Freiburg tradition (see, e.g., Streit 1995). For more details on common grounds and differences between Hayek and the early ordoliberals, see Streit and Wohlgemuth (1997), while Wohlgemuth (2013b) has more information on the overlapping between Austrian economics in general and ordoliberalism.

4.3 Ordoliberal Reference Model for European Integration

4.3.1 Importance of Individual Freedom

Like classical liberals and Austrian scholars, ordoliberals give priority to individual freedom in social, political and economic life (Streit 1995; Horn 2010; Butler 2015). For the early ordoliberals, this presumption was mainly derived from a Christian perspective as many of them (e.g. Eucken, Röpke and Müller-Armack) were practising Christians.[4] In this view, freedom is something given to every single person by God that constitutes his dignity. This means personal responsibility before God, respect for his fellow men as well as neighbourly love as the central driving force for voluntary solidarity (Schüller 2001). In addition, individual freedom is valued in a utilitarian sense following Jeremy Bentham (1746–1832), Adam Smith (1723–1790) and David Hume (1711–1776) as the best way to maximise the welfare of society as a whole, because it allows people to display their creativity and engage voluntarily in the mutually beneficial exchange of goods and services (Eucken 1952/1990, pp. 155ff.; Butler 2015).

Yet, all liberals agree that this individual freedom can never be absolute since different people's freedoms may conflict. So there has to be a minimum necessary restraint to protect the freedom of other individuals. For ordoliberals, competition is the most effective tool to constrain the misuse of individual freedom to the detriment of others, while at the same time preserving his personal dignity and unfolding people's productive power to the fullest extent possible.

4.3.2 Central Role of Competition

Competition is central in the ordoliberal approach as an instrument of constraining power, but also as a discovery procedure. It should therefore

[4] For more detail, see Goldschmidt (1998), Lorch (2013), Plickert (2016), and Hien (2017).

be at the core of European integration. The concern to establish a mechanism that would effectively constrain power was central to early ordoliberal thinkers (see Böhm 1957/1960) and motivated by German economic and legal history where private attempts to close markets, e.g. by forming cartels, were considered legitimate uses of the freedom of contract. If economic power becomes vested in the formation of private law, it impairs the political system and allows for infringements on the liberty of others (Streit and Wohlgemuth 1997). At the same time, it cripples the price mechanism and its allocative potential. Competition curtails economic as well as political power and so safeguards individual freedom by giving the opportunity to choose and run away from bad to better transaction partners. That is why Böhm (1961, p. 22) called competition "the most genial disempowering instrument".

Hayek (1978) highlighted competition's function as a discovery procedure to overcome the problem of knowledge. As he (1945) explains, the knowledge of what is needed, who needs it, and who has the means to meet these needs is dispersed and fragmented among the millions of individuals who compose society and is often held in inarticulate forms. Moreover, the cognitive abilities of every human being to capture and process all of this scattered information are limited. This makes it impossible for a centralised body of experts and politicians to gather in its totality the knowledge required to steer an economy into a certain direction. Instead, market competition through profit-and-loss feedbacks and changes in relative prices is best suited to mobilise the available bits of scattered knowledge to ensure a high level of prosperity. Profit-and-loss feedbacks provide the necessary incentive for individuals to acquire constantly new knowledge about consumer needs and how these should be met. If they do it successfully, they are rewarded by profits; otherwise, they suffer losses which urge them to correct their errors. Meanwhile, Kirzner (1973, 1997) emphasised that the main driving force in this discovery procedure is not governments but alert private entrepreneurs. Because entrepreneurs invest their own resources, they must be careful and astute when making investment decisions, while government bodies are generally institutionally precluded from capturing pecuniary profits in the course of their activities (Kirzner 1978).

4.3.3 Institutional Requirements

To unleash and maintain competition in both its disempowering and discovering functions, certain institutional requirements must be met. They are most prominently described by the famous "principles" laid down by Eucken (1952/1990, pp. 254ff.). Although they have been developed for a national economy, they can also be applied to supranational entities (Bökenkamp and Hesse 2016). At the core of the so-called constituent principles, conducive to the emergence and maintenance of the competitive order, is a workable price system that in the Hayekian interpretation transmits information in a codified manner about what should be produced and supplied and how this should be done and when. Further necessary principles are sound money, open markets, freedom to contract, private property, the enforcement of accountability (*Haftung*) for one's actions and predictability of economic policy. If competition remains hampered despite the implementation of these principles, Eucken (1952a, p. 292ff.) proposes further political interventions which he laid down in the so-called regulating principles. These comprise antitrust laws, income redistributing tax laws and laws to curb or avoid external effects.

While most Austrian economists question the need for these regulating interventions, they as well as classical liberals and institutional economists widely accept the importance of the constituent principles (Butler 2010, 2015). Hayek (1973, 1960) argued that a predictable, non-violent and self-regulating competitive and social order arose only when these principles were general (without exceptions), universal (applying to everyone) and stable (not changing very often). Such an order then tackles economic challenges with greater creativity and effectiveness than any centrally planned order.

4.3.4 Restraining Political Power

Conferring upon the state, the task of creating and ensuring a functioning institutional framework for the competitive order poses the problem of the abuse of political power. Eucken (1932) and Böhm (1933) saw politicians in much the same way they viewed economic agents: as self-interested actors rather than benevolent social wealth maximisers exclusively committed to the public interest. Policy-makers are not only exposed to the pressure of interest groups seeking selective benefits that provide them advantages over rival firms (e.g. subsidies and protection against competitors). They have an incentive to supply rent to secure re-election and retain power (Buchanan 1987). This is why the Freiburg ordoliberals called for a strong but limited state. It should be strong enough to resist the pressure of interest groups, but at the same time limited to pursue only the genuine task of protecting competition using market-compatible instruments (*Ordnungspolitik*).

However, the early ordoliberals grouped around Eucken did not elaborate on how a political constitution could look like to prevent and limit the arbitrary use of political power.[5] Later, Hayek (1979, ch. 17) did develop such a constitutional proposal. He argued that the legislative institutions of the time were preoccupied with awarding privileges to interest groups to acquire political support to the detriment of the general rule. Hayek therefore proposed a bicameral legislature: an upper chamber (the legislative assembly, pp. 112ff.) limited to enacting abstract rules needed to preserve the competitive order, and a lower chamber (the governmental assembly; pp. 119ff.) controlling the government and deciding on the provision of public services.

At least formally, the EU has a bicameral legislature, with the Council of Ministers as a sort of upper chamber and the European Parliament as the lower chamber. However, although its powers have continuously increased, the European Parliament neither has the power nor the will to repeal acts of the Commission. As *The Economist* (2017) has reported, rather than controlling spending and curbing the executive, the European Parliament often behaves more as a lobby group "whose main aim seems to be to spend more and to augment its own powers". As a result, the national heads of government, who make important decisions

[5]As Vanberg (1988, p. 24) put it, ordoliberals have "devoted astoundingly little explicit argument to the issue of what constitutional provisions might be required to make the 'strong government' perform to its proper task and not use its power in an undesired way".

in the European Council in a bargaining process, can effectively ignore their parliaments and bypass national voters (see also Streit 2011).

Given the peculiarities of the democratic process, attempts to profoundly change the existing rules of the game face insurmountable difficulties. This would require a binding decision of self-restraint by those who neither have a vested interest nor are likely to face massive pressure from the voters to do so (Streit 2000). In this situation, the only effective way to curb political power is competition between different jurisdictions. This allows people to choose not only through voting but also through exiting. At the same time, competition between legal entities acts as a discovery procedure to find out the appropriate institutional framework for a market order. The exit of mobile resources (capital and labour) allows citizens to test the expediency of available institutional arrangements and induces political actors to adapt to the preferences of the population and to develop innovative institutional solutions to the existing problems (see also Streit and Wohlgemuth 1997). Therefore, interjurisdictional competition must be an essential element of the ordoliberal concept of the EU.

4.3.5 Decentralisation

This insight into the disempowering and discovery properties of competition prompts a further normative conclusion for European integration: that political decision-making should be decentralised as much as possible to the national or even regional and local levels following the principle of subsidiarity (e.g. Hayek 1944/2006, 1979; Röpke 1951/2009, 1959, 1961/1964). This not only alleviates the power and knowledge problem, as political decision-makers at lower levels have a better understanding of local conditions, but also better ensures respect for national peculiarities and preferences.

Ordoliberal scholars fear that a powerful supranational authority in Europe might misuse its power to impose interventionist and redistributive policies between regions and so limit economic freedom. In this context, they stress the advantage of small political entities. This is in contrast to the view popular among proponents of an ever closer union that single

European countries are too small and weak to compete effectively with big global players such as the USA, Russia or China. However, Röpke (1959, p. 170f.) pointed out that such big entities are more tempted to harmonise and protect their economies and so limit freedom, whereas for small states closing the economy is not an option. Therefore, small states are "islands of economic reason" (Röpke 1959, p. 170; Mises 1927/2002; Marquardt and Bagus 2017; Vaubel 2017). Raico (2013) underscores in his analysis of the origins of the "European miracle" that Europe's economic success in the past few centuries has essentially been rooted in its diversity and the coexistence of and competition between small states, which creates the ability for its people to easily emigrate from one political jurisdiction to another and learn from each other.[6]

4.3.6 The Right to Exit

The disempowering and discovery functions of competition can further be enhanced in a supranational association if not only the individual citizens have the right to exit but also its subunits. This implies as an additional normative conclusion their right to secession. As a number of studies show,[7] federal states tend to centralise, mainly for two reasons. First, politicians and bureaucrats aim to increase their power by establishing tax and regulatory cartels, and second, bureaucrats and organised interest groups try to escape the attention of voters by shifting

---

[6]British philosopher and political economist John Stuart Mill (1806–1873) and Germany's greatest poet Johann Wolfgang von Goethe (1749–1832) argued in much the same vein. Mill (1869/1999, Ch. 3) concluded: "What has made the European family of nations an improving, instead of a stationary portion of mankind? Not any superior excellence in them, which when it exists, exists as the effect, not as the cause; but their remarkable diversity of character and culture. Individuals, classes, nations, have been extremely unlike one another…Europe is, in my judgment, wholly indebted to this plurality of paths for its progressive and many-sided development". Goethe, meanwhile, wrote in 1828: "*Frankfurt, Bremen, Hamburg, Lübeck sind groß und glänzend, ihre Wirkungen auf den Wohlstand von Deutschland gar nicht zu berechnen. Würden sie aber wohl bleiben, was sie sind, wenn sie ihre eigene Souveränität verlieren und irgendeinem großen deutschen Reich als Provinzialstädte einverleibt werden sollten? – Ich habe Ursache, daran zu zweifeln*" (in: Johann Peter Eckerman: *Gespräche mit Goethe in den letzten Jahren seines Lebens* - Kapitel 287, http://gutenberg.spiegel.de/buch/-1912/287).

[7]See e.g. Krane (1988), Vaubel (1994, 1996, 2009).

political decision-making away from the local to the central or supranational level (Vaubel 2013). In this situation, "secession, or the threat thereof, represents the only means through which the ultimate powers of the central government might be held in check. Absent the secession prospect, the federal government may, by overstepping its constitutionally assigned limits, extract surplus value from the citizenry almost at will, because there would exist no effective means of escape" (Buchanan 1995/2001, p. 71). Thus, secession is seen as a defence right against centralistic and discriminatory legislation that, if formally enshrined in the constitution or treaties, puts politicians and bureaucrats of different countries under competitive pressure to provide less centralisation, redistribution and other discriminatory legislation (Doering 2002; Vaubel 2013). The right to secession has in fact been incorporated in the Lisbon Treaty where Article 50 gives member states the right to leave the EU. Yet, hardly anyone in the EU institutions expected that a member country might really choose this option as they do everything they can to discourage secession (Vaubel 2013).

4.3.7 A Flexible EU of Different Clubs

Table 4.1 summarises the cornerstones of the ordoliberal reference model for European integration. Setting them implies a rather loose and flexible cooperation of states as the desirable form of European integration that respects national preferences with a high degree of decentralisation based on free economic and institutional competition including the right to exit the association as a whole, as well as certain fields in which member states want to do less together. The result would thus be a community of different clubs based on the common market as its core. The former liberal European Commissioner with dual German-British citizenship, Ralf Dahrendorf (1929–2009), called this in 1979 a "Europe à la carte". Solidarity in this model of Europe is not equal to large transfer payments among member countries, because they eliminate the incentives for market reforms and in the long run create a situation where some countries live at the expense of others, which, in turn, raises resentment among the member countries. On the

Table 4.1 Cornerstones of an ordoliberal reference model for European integration

Building blocks	Purpose
Individual freedom	Value in its own based on Christianity Driving force of economic progress
Competition • Within the national economies • Among member states • Between member and non-member states	Dismantle economic and political power Discovery procedure • In the economic sphere • In the political sphere (institutional competition)
Institutional requirements • Constituent principles • Regulating principles • Universalisability	Keeping markets open
Restrained political power Decentralisation and right to exit	Prevent rent seeking and provision Restrain economic and political power Reducing epistemological problems

Source Own depiction

contrary, European solidarity in this sense is playing according to the rules of market competition, which includes the application of the principle of accountability for one's decisions (Erhard 1957/1964; Röpke 1961/1964).

This ordoliberal reference model of Europe has the advantage of withstanding the strains wrought by human imperfections which arise from the lack of omniscience (limited knowledge) and benevolence (self-interested behaviour) of both economic and political decision-makers. The competitive order can correct for these imperfections by allowing a competitive trial and error learning process, allowing freedom of entry and exit, and by channelling selfish motives of individuals into what is beneficial to the society. The central problem of the eurozone and the migration crises stems exactly from the self-interested behaviour of its member countries that violate (d) agreed-upon rules (of the Stability and Growth Pact and the Dublin Accord). This, in turn, is related to the missing credibility to impose sanctions on those who break rules. Hence, the fewer areas are regulated by integration from above the fewer opportunities to violate rules. In addition, the social interdependence costs understood as the sum of decision-making costs and external costs will be much lower than in a one-size-fits-all ever closer union. In a Union of 27 diverse member states, decision-making costs are high because it is more difficult to agree in a heterogeneous than a homogeneous group. Equally, the external cost resulting from the choices made by the relevant body contrary to the individual country's own interest rises with the fear of being overruled in majority voting (Buchanan and Tullock 1962).

4.4 Scenarios for EU's Post-Brexit Economic Order

Documents as part of the White Paper process provide indications about the possible shape of the EU's economic order post-Brexit. The White Paper process was initiated by the European Commission led by Jean-Claude Juncker in response to the Brexit vote, and it sketches five scenarios of what the EU27 could look like by 2025.[8] The documents address the order among the remaining member states as well as the relations to the UK and other countries outside the EU.

4.4.1 The Internal Order of the EU

The five scenarios of how the EU's internal economic governance could develop in the next decade range from business as usual to a "multi-speed" Europe to "doing less more efficiently" and "doing much more together".

From the ordoliberal perspective, the preferable option would be scenario 2 (nothing but the single market), complemented with elements of scenario 3 (those who want more do more) and scenario 4 (doing less more efficiently), as this comes nearest to the reference model of a loose, decentralised and flexible association based on economic and

[8]All the documents of the White Paper process are accessible at https://ec.europa.eu/commission/white-paper-future-europe-reflections-and-scenarios-eu27_en.

institutional competition. However, in the last two scenarios, some problems are caused by areas that the Commission suggests should be assigned to the supranational level. These include the harmonisation of taxes, social, environmental and consumer protection standards, and innovation policy. This would be detrimental to ordoliberal principles, because it further reduces and, in the extreme case, eliminates economic and institutional competition as a discovery and disempowering procedure.

A further problem with option 3 is that it does not specify what "those who want more do more" exactly means. In today's EU, there are already coalitions of those wanting to do more together in the Schengen and euro areas. Widening this option to cover other areas, such as defence and social issues, would be in line with the ordoliberal concept of flexibility. However, this would not be the case if this suggestion were to be understood as an intermediate stage on a predefined trajectory towards an ever closer union where some member states move ahead faster than others. Such an "ever-closer-one-size-fits-all-sooner-or-later" option is not really compatible with genuine flexibility, where different countries engage in mutual integration in different policy areas (Wohlgemuth 2017c).

Although officially neutral, the European Commission made it clear that it was least enthusiastic about option 2 (Spiegel-Online 2017), while favouring the most federal option of "doing much more together" (Wohlgemuth 2017b; Berthold 2017). As expressed in the White Paper, the Commission believes that a big disadvantage of the second scenario is the persistence of different tax policies as well as product, social and environmental standards, because it fears a "race to the bottom". Furthermore, the Commission worries about the resurgence of bilateralisms that might prevent the EU from speaking with one voice in a number of international fora on global issues such as "climate change, fighting tax evasion, harnessing globalisation and promoting international trade". This belief is in stark contrast to the ordoliberal insights as it fails to understand the productivity and freedom-enhancing powers of institutional competition. As Kirzner (1973, 1997) pointed out, what is crucial ("necessary and sufficient") to induce and maintain competition is that markets are kept open. But this does not require harmonisation of all national rules but their mutual recognition following the Cassis de Dijon decision of 1979. This implies that products and services lawfully produced in one member state must be freely marketable in all other member countries. Hence, based on these institutional prerequisites, ordoliberals unequivocally support the free flow of products, services and capital. With regard to the free movement of labour, Röpke (1959) would have shared the Brexiteers' demand for an encompassing right of the national states to steer external and internal EU immigration to protect national identity and not to overburden the accommodation capacities of a country to integrate migrants.

European Commission President Jean-Claude Juncker reaffirmed his preference for the great leap forward towards an ever closer federal state in his State of the Union Address on 13 September 2017 when he proposed an expansion of the Schengen area, the eurozone and the banking union to the whole of the European Union as well as the establishment of a European Social Standards Union. This would amount to pressing ahead with a one-size-fits-all policy that furthers economic and institutional competition as a discovery and disempowering procedure. In doing so, the EU would undermine the foundation for liberty and prosperity. According to Hayek (1958), a free common market is a sufficient basis for a working interstate federation as it prevents government meddling with the economy and preserves peace by keeping social interdependency costs low.

4.4.2 Relations with Non-EU Members

Equally problematic from an ordoliberal perspective are suggestions outlined in the new reflection paper on globalisation as well as in Juncker's State of the Union address on external relations with non-EU countries. Both documents underline the EU's reluctance to make unilateral concessions and grant free trade with non-EU members at any price. The EU wants to defend its "strategic interest" and "legitimate public policy objectives" and "fight for a fairer global order". In this context, "fair" means provision of "a level playing field" by preventing or eliminating tax evasion, government subsidies and "social dumping" through "harmful and unfair" lower social, health and environmental standards" (p. 13). Instead, the EU wants to promote a "'race-to-the top' approach" (p. 13). Rather than fostering economic and institutional competition to curb excessive taxation and to find appropriate standards as well as other institutional settings and innovations that meet consumers' preferences, this policy is protectionist in nature as it aims at raising the rivals' cost and making all countries equally sclerotic and uncompetitive. Hence, the policy conflicts with the crucial ordoliberal constituent principle of open markets.

Keeping markets open through eliminating entry barriers is essential for enhancing the disempowering and discovery properties of competition. Therefore, ordoliberals would suggest upholding and deepening the common market with the UK after Brexit and even recommending unilateral trade liberalisation. However, for political reasons, this seems a quite unlikely option for the EU27. Not only would this conflict with the EU's foreign trade policy stance, which is predicated on the reciprocity of intergovernmental bargains. In the EU institutions, there are also strong forces that want to deter other EU countries with large Euro-sceptic movements from choosing the same path as the UK.[9] Therefore, a new institutional arrangement for the EU27-UK relations will have to be negotiated.

4.5 What Drives Germany's Post-Brexit EU Policy?

Regardless of whether Germany favours a certain scenario or even a distinct direction for EU policy post-Brexit, the EU economic order presupposes that German policy-makers are guided by a specific economic philosophy and that there is a realistic chance to implement it.

---

[9]See, e.g., a BBC report (http://www.bbc.com/news/uk-politics-41140564) citing the EU's chief Brexit negotiator Michel Barnier as saying: "We intend to teach people… what leaving the single market means".

4.5.1 Economic Philosophy

In the ongoing public debate on Germany's role in the EU, scholars and media have focused on the German government's appeal to obey rules and undertake austerity measures in the management of the eurozone crisis—amid calls by other member countries for transfers and Keynesian policies—in a clear reference to the legacy of ordoliberalism, the economic philosophy underlying Germany's social market economy concept (see, e.g., Bonefeld 2012; Dullien and Guerot 2012; Blyth 2013; Biebricher 2014; Van Esch 2014; Young 2014; *Economist* 2015; Nedergaard and Snaith 2015; Stelzenmüller 2015; Feld et al. 2015; Brunnermeier et al. 2016; Cardwell and Snaith 2018). Therefore, one might expect that the potential ordoliberal governance of the EU's economic system will be strengthened after Brexit. However, this presupposes that the German policy-makers' mindset is really shaped by that economic strain of thinking.

To what extent, if at all, ordoliberalism drives German policy-makers is a complex question that faces insurmountable difficulties in disentangling various influences (see also Biebricher 2014). A content analysis of speeches by selected political figures in terms of how often they refer to ordoliberal terms and thinkers, as conducted by Hien (2017), is at best a first hint. Even if politicians publicly referred to ordoliberal principles, actual economic policy might point to the opposite.

Even though the rules of the game are central to ordoliberal thinking, and the insistence on accountability for one's debts and on low inflation reflects two of Eucken's constituent principles, accountability and good money, they alone do not make German economic policy ordoliberal. Meanwhile, Germany's inclination to rules may also reflect a historical experience that can be traced back to the Holy Roman Empire of the German Nation. Like today's EU, it was a multi-ethnic complex composed of hundreds of quasi-independent subunits (principalities and duchies). Power was highly decentralised and vested in these entities. Yet, what bound them loosely together under a relatively powerless emperor was rules that sanctioned these rights and regulated the relation to the emperor. The same holds for the constituent states of the German Empire after 1871. As Leipold (2006) explains, these legal bonds and regulations were crucial in shaping Germany's informal institutional setting in the century-long absence of common ideological and religious ties and a common national identity (see also Habermann 2013).

Likewise, the preference for low inflation could just as well be explained by Germany's bad historic experience with the hyperinflation of 1923 and 1945 (Bökenkamp 2016). And Germany's call to curb budget deficits could merely be an expression of its national self-interest for the simple politico-economic reason of limiting Germany's liability for other countries' wrong policies and of avoiding the moral hazard arising from easy access to someone else's money. Although a large section of the German public is pro-European, it might be difficult to explain to domestic voters the need for constant financial transfers to countries that are reluctant to correct flawed policies.

Moreover, an ordoliberal economic policy agenda would require that not just two, but *all* constituent principles are fulfilled. So insisting on low inflation and debts while fixing prices in certain markets would be inconsistent and not compatible with ordoliberalism. Furthermore, ordoliberalism is not about setting and enforcing rules per se. It is about setting and enforcing *rules that are conducive to competition* as a discovering and disempowering procedure. This is an essential difference as rules or laws can also be set and enforced to restrict or even oust competition. In this respect, too, Germany has a long tradition that goes back to eighteenth-century Prussia and the concept of the police and welfare state. Characteristic for its style was a strong omnipresent government that was not only responsible for the enforcement of law and order, but also for the social well-being of its citizens and therefore regulated through "enlightened" bureaucrats and kings (for more detail, see Habermann 2013; Leipold 2006). As a result, until the post-World War II period, Germany actually had a statist tradition (Habermann 2013; Wehler 2014; Bökenkamp 2016; Brunnermeier et al. 2016). By contrast, France until World War II advocated laissez-faire and a rule-based economic policy referring to classical liberal economists Jean-Baptiste Say and Frédéric Bastiat (Bökenkamp 2016; Brunnermeier et al. 2016).

But German economic policy after the end of World War II, specifically the country's social market economy system, was also full of examples of rules that hampered competition and favoured particular vested interests. One of the latest examples is a renewable energy policy measure—instituted by the new federal government of Chancellor Angela Merkel—that completely banned market forces. A similar policy scheme was being discussed with regard to the automotive industry under the motto *Verkehrswende* to promote e-cars (see, e.g., Gastel 2017). The level of subsidies to the selected sectors such as coal, steel, shipbuilding, energy and transport rose almost uninterruptedly beginning in 1970, even after Chancellor Kohl assumed office and his conservative-liberal coalition government committed itself to reducing such subsidies and strengthening market forces. According to Laaser and Rosenschon (2016), in 2015, the total volume of subsidies reached its highest level ever, at EUR 168 billion. Other more recent violations of ordoliberal principles are the minimum wages and maximum prices in the rental market—because they are government interventions into the market process in pursuit of specific market outcomes (see also Erlei 2014).

Ordoliberalism is considered to have been most influential only in the first phase of the social market economy from 1948 to 1966 under Ludwig Erhard, Germany's first post-war economics minister until 1963 and later chancellor until 1966 (Habermann 2013; Sally 2016). Although he made the notion of the social market economy popular, Erhard clearly stood for a free market economic policy on both the national and European levels. For Erhard, the social market economy was not understood as a "third way" between extreme socialism and extreme capitalism, but a market economy (Goldschmidt 2004). He (1966, p. 320) was convinced that "the freer an economy is, the more social it is". Ludwig Erhard's free market policy worked and transformed West Germany into Europe's economic powerhouse. However, Erhard and his ordoliberal advisors did not succeed in making the reforms long-lasting (see Rüstow 1961; Röpke 1966; Habermann 2013). Many far-reaching regulations of markets and industries prevailed (e.g. in agriculture, housing, transportation, energy, services and crafts), and antitrust legislation was filled with many exemption clauses that allowed for the restriction of competition.

In the 1960s, the ordoliberal component of economic policy progressively diminished with the rise of Keynesianism and the persistence of egalitarian ideas (Feld et al. 2015). As a result, socially and, in recent times, ecologically motivated interventions increased and gradually displaced the market (for more detail, see Schüller 2002; Bökenkamp 2010). Like in Prussia's police and welfare state, the view came to prevail that enlightened benevolent policy-makers must and can engineer the free market to produce desirable results in the name of social justice and ecology (Habermann 2013). This resulted in a regulated economy that is ranked 26th worldwide (down from 17th) in the latest 2017 Index of Economic Freedom compiled by the Heritage Foundation (2017), far behind Switzerland (No. 4), Estonia (No. 6) and the UK (No. 12), and in a vast welfare state that is increasingly ill-equipped to tackle demographic challenge. On the European level, since the Maastricht Treaty of 1992 Germany increasingly adopted the French approach to economic policy, with an emphasis on the "primacy of politics" and a preference for supranational collective actions (Schüller 2011). Chancellor Angela Merkel has continued this course. Her government has advocated an energy transition and tax harmonisation and for the most part supported France's push for a European economic government.

Ordoliberal principles seem to prevail at the Bundesbank and in the German Council of Economic Experts (GCEE), which advises the government and the country's central bank (Feld et al. 2015). In a lecture given in 2008, the former chief economist of the European Central Bank (ECB), Jürgen Stark (2008), said that the work of the ordoliberal Walter Eucken was "a constant source of inspiration" for him. In an annual Walter Eucken lecture in Freiburg in 2013, Jens Weidmann (2013) underscored the importance of ordoliberal insights in successfully managing the EU's manifold crises. Meanwhile, former GCEE Chairman Olaf Sievert (2003) said that the Council always focused on ordoliberal rather than interventionist policies. In the current line-up of the Council, one of its members, Lars Feld, concurrently works as the head of the Freiburg-based Walter Eucken Institute. Together with three of the four other members (except pro-unionist Keynesian economist Peter Bofinger), Feld is usually critical of the government for its interventionism.[10] However, their influence on politicians should not

[10]See, e.g., Jahresgutachten 2014/2015: "Mehr Vertrauen in Marktprozesse", or Jahresgutachten 2013/2014: "Gegen eine rückwärtsgewandte Wirtschaftspolitik".

be overestimated; decision-makers often take such advice with reservation or ignore it altogether (see Straubhaar 2014; *FAZ* 2009). So ordoliberal principles at best play only a minor role for Germany's government in economic policy-making. Therefore, for the time being, it is unlikely that Germany will be promoting a distinct ordoliberal alternative option for the EU's economic order after Brexit.

4.5.2 The Primacy of European Unity and Stability

European integration takes high priority in German politics. In view of the country's history, there is a deep-seated belief in Germany in the European project. Even in the preamble to its constitution, Germany committed itself to serve for peace as an equal partner in a united Europe. It is therefore almost taboo to question European integration (Wohlgemuth 2017a) and an imperative of Germany's foreign and European policies irrespective of party lines to keep the EU united and stable and continue integration. However, this always requires compromises to bridge differences between different members and strains of thinking (Steinmeier 2016; Besch and Odendahl 2017; Helwig 2017). Traditionally, in these efforts, Germany has always sought close coordination with France. Regardless of different policy preferences, these two countries usually succeeded in reaching comprises, making both countries major drivers of European integration. The introduction of the euro under Chancellor Helmut Kohl, despite warnings from many economists, was the most prominent example of the priority Germany gives to upholding European unity over economic principles.[11] This Franco-German axis, as "the saviour of the European project", has gained even more importance after the Brexit vote and the erosion of the hitherto strong German–Polish relationship since the Law and Justice party came to power in Poland in 2015. The Franco-German team-up gained new momentum after the election of Emmanuel Macron as France's new president on 14 May 2017.

[11]For more detail, see Bagus (2010), Starbatty (2013).

Macron won the election with a clear pro-EU agenda and concrete proposals, such as an economic government with a eurozone budget for joint investments and an EU finance minister. Although Germany is in general supportive of an economic government for the eurozone (see, e.g., *Welt* 2017), it is not clear to what extent both countries share the same understanding. Mussler (2011) and Wohlgemuth (2017a) argue that the current German government with Finance Minister Wolfgang Schäuble wants an economic government primarily to restore the rule-based economic coordination and surveillance of competitiveness in the eurozone, including the impartial automatic enforcement of the stability pact following strict criteria. But raising and spending European taxpayers' money or issuing joint eurobonds would not fall within the minister's remit (Wohlgemuth 2017c). Yet, this is only the position of the CDU and the liberal FDP. Ahead of the 24 September 2017 federal elections, the German liberals campaigned on a market-oriented message and called for phasing out bailout funds, an orderly state bankruptcy, a streamlined exit process in the eurozone (Helwig 2017) and an amicable Brexit (Wohlgemuth 2017d).

Instead of binding legal commitments, France favours political discretion and intergovernmental decisions on how to spend the money raised from common taxes or mutualised debt instruments. Yet, given the German government's interventionist economic policy record at home, it is likely that the visions of both the French and German governments for a reform of the eurozone might not differ that much (Berthold 2017). German ordoliberal economists often describe Angela Merkel as a chancellor with no principles, except for staying in power (Berthold 2017), and "changing her mind according to public opinion and political opportunity with disarming nonchalance" (Wohlgemuth 2017d). France's Macron, in his September 2017 "Initiative for Europe", and the European Commission, in its 6 December 2017 roadmap for deepening the eurozone, both called for a separate eurozone budget, a common EU finance minister, common deposit protection and the establishment of a European Monetary Fund. These proposals were in general welcomed by the new German federal government in its coalition treaty in March 2018 (Bundesregierung 2018). Meanwhile, in May 2018, a group of

154 economists, most of them anchored in the Freiburg and Austrian Schools as well as in Institutional Economics and Public Choice Theory, issued a public appeal to the new federal government in which they dismissed calls for deeper eurozone integration (*FAZ* 2018).

4.5.3 Germany's Position in EU Institutions

A further factor making it unlikely for Germany to push for a comprehensively distinct model for the EU's post-Brexit economic order is its position in key EU institutions where major decisions are made. This, in turn, depends on the voting rules.

Germany on its own can only block decisions in the permanent rescue-funding programme of the European Stability Mechanism (ESM) where decisions are made about austerity measures in exchange for loans. There, the voting rights of each ESM member are equal to the number of shares allocated to it in the authorised capital stock of the ESM as set out in a February 2012 annex to the treaty establishing the ESM. Germany's share in the paid-up capital is the largest, at almost 27%. It therefore has in principle a blocking minority. However, the most important decisions[12] taken by the EMS Board of Governors require mutual consent. Only in the ESM emergency voting procedure, which decides on financial assistance, and in areas of minor importance to the current anti-crisis policy,[13] can Germany use its blocking minority (European Stability Mechanism 2012).

In the European Central Bank, which was originally modelled after the German Bundesbank, Germany has even less power to influence monetary policy. Under the one-member-one-vote rule, Germany has regularly been outvoted in the ECB Governing Council by countries with a different view on macroeconomic stability (Italy, Portugal,

[12]These include decisions to provide stability support to an ESM member, the choice of instruments, conditions and terms of such support, calling in authorised unpaid capital, changing the authorised capital stock and adapting the maximum lending volume.

[13]They include setting out the detailed terms of accession of a new member to the ESM, appointing the managing director and approving the annual accounts of the ESM.

Greece, Spain and France). The same may happen with the proposed economic government for the eurozone. Even if it were to be modelled after fiscally responsible Germany, it is likely that Germany and its smaller allies, such as Austria, the Netherlands, Luxemburg, Finland, Ireland, the Baltic states and Slovakia, would find themselves in the minority amid efforts to push through the proposal (Trebesius 2017). Similarly, Germany has a minority position in the Supervisory Board and the Single Resolution Board of the banking union, because both these organisations follow the one-member-one-vote principle. However, the Council of the EU can reject decisions made by the Single Resolution Board on the liquidation of banks within 24 hours if it finds that the bankruptcy is against the public interest.

Nor in the EU Council can Germany alone impose its preferences in the qualified majority voting procedure despite its biggest share of votes by population (16.06%). It needs allies. Under Article 16 of the Treaty on the European Union, as of 1 November 2014 the qualified majority is reached when 55% of the member states (16 of 28, or 15 of 27 after Brexit) representing at least 65% of the EU's population approve an act. In the case of a proposal from neither the Commission nor the High Representative, the qualified majority is 72%. In addition, there is the possibility to stop proposals through a blocking minority. This must include at least four Council members representing at least 35% of the EU's population (European Council 2017).[14] Without the UK, voting power in the Council will shift towards statist-oriented rather than stability-minded countries. Germany will find it harder to build up blocking minorities.

Table 4.2 classifies EU member countries into liberal and statist economies, using the Heritage Foundation's 2017 Index of Economic Freedom, and assigns to them their votes by population according to an EU vote calculator. Although the more liberal-oriented countries will retain their blocking minority after Brexit, their share will decrease from almost 50% to 42%. The blocking minority might be lost, in particular

[14]In cases where not all member states participate in voting (e.g. acts adopted only by euro-area or Schengen member states, or within enhanced cooperation), the qualified majority is calculated only on the basis of the participating member states.

Table 4.2 Council voting weights in the EU before and after Brexit (% of total EU population)

Liberal economies			Statist economies		
Member state	Before Brexit	After Brexit	Member state	Before Brexit	After Brexit
EE	0.26	0.30	RO	3.87	4.44
IE	0.91	1.05	BG	1.40	1.61
UK	12.79	–	CY	0.17	0.19
LU	0.11	0.13	BE	2.21	2.53
NL	3.37	3.87	MT	0.09	0.10
LT	0.57	0.65	HU	1.92	2.21
DK	1.12	1.28	SK	1.06	1.21
SE	1.96	2.24	ES	9.09	10.42
LV	0.39	0.44	FR	13.05	14.96
FI	1.07	1.23	PT	2.02	2.32
DE	16.06	18.42	IT	12.00	13.76
CZ	2.04	2.34	HR	0.82	0.94
AT	1.71	1.96	SI	0.40	0.46
PL	7.43	8.52	EL	2.11	2.42
Total	*49.79*	*42.43*	*Total*	*50.21*	*57.57*

Source EU's vote calculator http://www.consilium.europa.eu/en/council-eu/voting-system/voting-calculator/; Heritage Foundation (2017)

if Poland, with its share of 8.52% of the population, moved from the liberal to the statist group. Such a move cannot be excluded given the Polish ruling Law and Justice party's interventionist and nationalist economic policies as well as its increasingly anti-German rhetoric. So this underscores the importance of the UK as an ally for promoting a distinct economic policy agenda.

In fact, the ordoliberal reference model of European integration described in Sect. 4.2 is largely congruent with the British vision of Europe and many Brexit arguments. While migration did play a role, some deeper causes are anchored in the British classical liberal tradition, which, despite periods of socialist experiments, was always present in Britain.[15] It has engrained the following principles of the British

[15]For a profound and detailed presentation of the Brexit arguments, see the film "Brexit: The Movie" (https://mises.org/blog/brexit-movie-makes-economic-case-against-eu) and Smith (2016).

mindset, which drove support for Brexit (Bökenkamp and Hesse 2016; Wirtz 2017): (1) free markets, (2) localism (i.e. policy-makers should be as close to citizens as possible) and (3) small government. In fact, according to one pollster, the second principle was the chief motive for "Leave" voters, both Tory and Labour (Ashcroft 2016).

As Bökenkamp and Hesse (2016) showed, all British prime ministers since the UK's accession to the European Economic Community (EEC) in 1973, irrespective of party affiliation, have viewed the EEC or EU not as an end in itself but as a means to enhance prosperity, freedom and democracy. Accordingly, free trade among member states *and* with the rest of the world is viewed as the heart of European integration, and the national states as its main actors. This implies subsidiarity, decentralisation and flexibility to accommodate the diversity of EU members. In contrast, for politicians in continental Europe, including former European Commission President Jacques Delors, the common market is a means to create a European federal state, and therefore, they advocate centralisation, harmonisation and regulation of the common market (Geddes 2013). Ludwig Erhard (e.g. 1957/1964) and Wilhelm Röpke (1959) were sympathetic to the British vision of Europe, while they heavily criticised the French view. So Britain would actually be Germany's most natural partner if it wanted to promote a profound ordoliberal reform path for the EU.

4.6 Conclusions

In the 1950s, Wilhelm Röpke anxiously argued against integration trajectories towards centralisation and harmonisation in Europe. He warned that such processes, instead of boosting economic growth and bringing people together, would prove to be an "explosive and an instrument of disintegration" (Röpke 1955, p. 96; similarly in Röpke [1959, §88]). Roughly 60 years later, the exit of the UK, the rise of Euro-sceptic movements and parties in many member countries and the infringement procedures against Poland are alarming indications in support of Röpke's warnings. So Europe is at a crossroads.

From the ordoliberal perspective, the way out of the current crises is not more unification, centralisation and harmonisation, but a flexible Europe of different clubs based on a free and open single market following the motto "those who will and can". This would best ensure competition as a disempowering and discovery procedure and at the same time account for Europe's heterogeneity. There are two rather spontaneous forces that might set the course in this direction contrary to the intentions of the European Commission and pro-ever-closer-union political elites. First, insurmountable collective action problems in an entity with 27 heterogeneous member states make it impossible to agree on further steps towards an ever closer union. Second, persistently slow growth resulting from the EU's institutional sclerosis exhausts its resources and leaves no other option than to "do less but more efficiently" and so to reinforce subsidiarity and rely on economic and institutional competition.

References

Acemoglu, D., & Robinson, J. (2012). *Why Nations Fail: The Origins of Power, Prosperity, and Poverty*. New York: Crown Business.

Ashcroft, L. (2016, June 25). *How the United Kingdom Voted on Thursday—And Why*. ConservativeHome. Available at http://www.conservativehome.com/platform/2016/06/lord-ashcroft-how-the-united-kingdom-voted-on-thursday-and-why.html.

Bagus, P. (2010). *The Tragedy of the Euro*. Auburn, AL: Ludwig von Mises Institute. Available at https://mises.org/library/tragedy-euro.

Berthold, N. (2017, June 30). Emmanuel Macron und Angela Merkel: Politische Lichtgestalt (?) und ordnungspolitische Geisterfahrerin (!). *Wirtschaftliche Freiheit*. Available at http://wirtschaftlichefreiheit.de/wordpress/?p=21063&print=1.

Besch, S., & Odendahl, C. (2017, March). *Berlin to the Rescue? A Closer Look at Germany's Position on Brexit*, Center for European Reform. Available at http://www.cer.eu/sites/default/files/pb_germany_brexit_17.3.17.pdf.

Biebricher, T. (2014). The Return of Ordoliberalism in Europe—Notes on a Research Agenda. *i-lex 9*, 1–24.

Blyth, M. (2013). *Austerity: The History of a Dangerous Idea*. Oxford: Oxford University Press.

Böhm, F. (1933). *Wettbewerb und Monopolkampf – Eine Untersuchung zur Frage des wirtschaftlichen Kampfrechtes und zur Frage der rechtlichen Struktur der geltenden Wirtschaftsordnung*. Berlin: Heymann.

Böhm, F. (1957/1960), Die Forschungs- und Lehrgemeinschaft zwischen Juristen und Volkswirten an der Universität Freiburg in den dreißiger und vierziger Jahren des 20. Jahrhunderts. In F. Böhm (Ed.), *Reden und Schriften* (pp. 158–175). Karlsruhe: Müller.

Böhm, F. (1961). Demokratie und ökonomische Macht. In Institut für ausländisches und internationales Wirtschaftsrecht (Ed.), *Kartelle und Monopole im modernen Recht* (pp. 1–24). Karlsruhe: Müller.

Bökenkamp, G. (2010). *Das Ende des Wirtschaftswunder: Geschichte der Sozial-, Wirtschafts- und Finanzpolitik in der Bundesrepublik 1969–1998*. Stuttgart: Lucius & Lucius.

Bökenkamp, G. (2016, September 27). *Kulturgeschichte: Liberales Frankreich, etatistisches Deutschland?* Open Europe Berlin. Available at http://www.openeuropeberlin.de/kulturgeschichte-liberales-frankreich-etatistisches-deutschland-wie-die-zwei-nachbarn-die-rollen-tauschten-und-was-das-fuer-die-gegenwart-bedeutet-von-gerard-boekenkamp/.

Bökenkamp, G., & Hesse, N. C. (2016). *Grossbritannien, die EU und die liberale Vision für Europa: Zwischen Brexit, Politischer Union und Reform* (Friedrich A. von Hayek-Gesellschaft). Norderstedt: Books on Demand.

Bonefeld, W. (2012). Freedom and the Strong State: On German Ordoliberalism. *New Political Economy, 7*(5), 633–656.

Brunnermeier, M. K., James, H., & Landau, J.-P. (2016). *Euro and the Battle of Ideas*. Princeton, NJ: Princeton University Press.

Buchanan, J. M. (1987). The Constitution of Economic Policy. *The American Economic Review, 77*(3), 243–250.

Buchanan, J. M. (1995/2001). Federalism as an Ideal Political Order and an Objective for Constitutional Reform. In *The Collected Works of James M. Buchanan* (pp. 67–78). Indianapolis: Liberty Fund.

Buchanan, J. M., & Tullock, G. (1962). *The Calculus of Consent*. Ann Arbor: University of Michigan Press.

Bundesregierung. (2018). *Koalitionsvertrag 2018*. Available at https://www.bundesregierung.de/Content/DE/_Anlagen/2018/03/2018-03-14-koalitionsvertrag.html.

Butler, E. (2010). *Austrian Economics: A Primer*. London: Adam Smith Institute.

Butler, E. (2015). *Classical Liberalism: A Primer*. London: Institute of Economic Affairs.

Cardwell, P. J., & Snaith, H. (2018). 'There's a Brand New Talk, But It's Not Very Clear': Can the Contemporary EU Really Be Characterized as Ordoliberal? *Journal of Common Market Studies, 56*(5), 1053–1069.

Dahrendorf, R. (1979). *A Third Europe? Third Jean Monnet Lecture*. European University Institute, 26 November 1979. Available at http://aei.pitt.edu/11346/2/11346.pdf.

Doering, D. (2002). *Friedlicher Austritt. Braucht die Europäische Union ein Sezessionsrecht?* Available at http://www.progress-foundation.ch/de/search.

Dullien, S., & Guérot, U. (2012). *The Long Shadow of Ordoliberalism: Germany's Approach to the Euro Crisis* (Policy Brief ECFR/49). European Council on Foreign Relations. Available at http://www.ecfr.eu/page/-/ECFR49_GERMANY_BRIEF.pdf.

Economist. (2015, May 9). Of Rules and Order. German Ordoliberalism Has Had a Big Influence on Policy During the Euro Crisis. http://www.economist.com/news/europe/21650565-german-ordoliberalism-has-had-big-influence-policy-during-euro-crisis-rules-and-order.

Eder, F., Heath, R., Barigazzi, J., & Ariès, Q. (2017, August 17). Commission Outlines 5 Scenarios for Future of EU in White Paper. *Politico*. http://www.politico.eu/article/breaking-politico-obtains-white-paper-on-eu-future/.

Erhard, L. (1957/1964). *Wohlstand für alle*. Düsseldorf: Econ-Verlag [hier: 8th ed., edited by Wolfram Langer (1964), Ludwig-Erhard-Stiftung e.V., Bonn].

Erhard, L. (1966). *Wirken und Reden*. Ludwigsburg: Verlag Hoch.

Erlei, M. (2014, June 24). Die Institutionen der Sozialen Marktwirtschaft erodieren mit steigender Geschwindigkeit. *Wirtschaftliche Freiheit*. Available at http://wirtschaftlichefreiheit.de/wordpress/?p=15101.

Eucken, W. (1932). Staatliche Strkturwandlungen und die Krise und die Krise des Kapitalismus. *Weltwirtschaftliches Archiv, 84*, 297–331.

Eucken, W. (1952). *Grundsätze der Wirtschaftspolitik* (New edition 1990). Tübingen: Mohr Siebeck.

European Council. (2017). *Voting System*. Available at http://www.consilium.europa.eu/en/council-eu/voting-system/.

European Stability Mechanism. (2012). *ESM Treaty—Consolidated Version*. Available at http://www.esm.europa.eu/pdf/ESM%20Treaty%20consolidated%2013-03-2014.pdf.

FAZ—Frankfurter Allgemeine Zeitung. (2009, November 13). Kritik der fünf Weisen lässt Merkel kalt. Available at http://www.faz.net/aktuell/wirtschaft/wirtschaftspolitik/gutachten-vorgelegt-kritik-der-fuenf-weisen-laesst-merkel-kalt-1882830.html.

FAZ—Frankfurter Allgemeine Zeitung. (2018, May 21). Aufruf: Der Euro darf nicht in die Haftungsunion führen. Available at http://www.faz.net/aktuell/wirtschaft/eurokrise/oekonomen-aufruf-euro-darf-nicht-in-haftungsunion-fuehren-15600325.html.

Feld, L. P., Köhler, E. A., & Nientiedt, D. (2015). *Ordoliberalism, Pragmatism and the Eurozone Crisis: How the German Tradition Shaped Economic Policy in Europe* (Freiburg Discussion Papers on Constitutional Economics, 2015-4). Freiburg: Walter Eucken Institut.

Gastel, M. (2017, June 28). Mit der Verkehrswende unsere Mobilität sichern. *Zeit-Online*. Available at http://www.zeit.de/politik/deutschland/2017-06/verkehrswende-gruene-elektromobilitaet-klimawandel-zukunft.

Geddes, A. (2013). *Britain and the European Union*. Basingstoke: Palgrave Macmillan.

Goldschmidt, N. (1998). Christlicher Glaube, Wirtschaftstheorie und Praxisbezug. Walter Eucken und die Anlage 4 der Denkschrift des Freiburger Bonhoeffer-Kreises. *Historisch-Politische Mitteilungen, 5*(1), 33–48. Available at http://www.kas.de/upload/ACDP/HPM/HPM_05_98/HPM_05_98_2.pdf.

Goldschmidt, N. (2004). *Alfred Müller-Armack and Ludwig Erhard: Social Market Liberalis* (Freiburg Discussion Papers on Constitutional Economics, No. 04/12). Available at http://www.econstor.eu/bitstream/10419/4344/1/04_12bw.pdf.

Goldschmidt, N. (2006). Vom Nutzen der Ordnungsökonomik angesichts einer mathematisierten Wirtschaftswissenschaft. *Wirtschaftsdienst-Zeitschrift für Wirtschaftspolitik, 86*(1), 14–18.

Goldschmidt, N., Wegner, G., Wohlgemuth, M., & Zweynert, J. (2009, June 19). Was ist und was kann Ordnungsökonomik? *FAZ—Frankfurter Allgemeine Zeitung*. Available at http://www.faz.net/aktuell/wirtschaft/wirtschaftswissen/wissenschaftsdebatte-was-ist-und-was-kann-ordnungsoekonomik-1818053.html.

Habermann, G. (2013). *Der Wohlfahrtsstaat: Ende einer Illusion*. München: FinanzBuch Verlag.

Hayek, F. A. (1944/2006). *The Road to Serfdom*. London and New York: Routledge.

Hayek, F. A. (1945). The Use of Knowledge in Society. *The American Economic Review, 35*(4), 519–530.

Hayek, F. A. (1958). The Economic Conditions of Interstate Federalism. In F. A. Hayek (Ed.), *Individualism and Economic Order* (pp. 255–272). Chicago: University of Chicago Press.

Hayek, F. A. (1960). *The Constitution of Liberty*. Chicago: University of Chicago Press.

Hayek, F. A. (1973). *Law, Legislation and Liberty, Volume 1: Rules and Order*. Chicago: University of Chicago Press.

Hayek, F. A. (1978). Competition as a Discovery Procedure. In F. A. Hayek (Ed.), *New Studies in Philosophy, Politics, and Economics and the History of Ideas* (pp. 179–190). Chicago: University of Chicago Press.

Hayek, F. A. (1979). *Law, Legislation and Liberty, Vol. III: The Political Order of a Free People*. Chicago: University of Chicago Press.

Helwig, N. (2017, August). *The EU in the German Elections* (Fiia Briefing Paper 225). The Finnish Institute of International Affair. Available at http://www.fiia.fi/en/publication/703/the_eu_in_the_german_elections/.

Heritage Foundation. (2017). *Index of Economic Freedom 2017*. Available at http://www.heritage.org/index/download.

Hien, J. (2017, September 10). The Religious Foundations of the European Crisis. *Journal of Common Market Studies*, OnlineFirst. https://doi.org/10.1111/jcms.12635.

Horn, K. I. (2010). *Die Soziale Marktwirtschaft: Alles, was Sie über den Neoliberalismus wis-sen sollten*. Frankfurt: Frankfurter Allgemeine Buch.

Kirzner, I. M. (1973). *Competition and Entrepreneurship*. Chicago: University Chicago Press.

Kirzner, I. M. (1978). *The Perils of Regulation: A Market-Process Approach*. Coral Gables, FL: University of Miami, School of Law.

Kirzner, I. M. (1997, March). Entrepreneurial Discovery and the Competitive Market Process: An Austrian Approach. *Journal of Economic Literature*, *XXV*, 60–85.

Krane, D. A. (1988). The Evolutionary Pattern of Federal States. In C. L. Brown-John (Ed.), *Centralizing and Decentralizing Trends in Federal States* (pp. 39–62). Lanham, MD: University Press of America.

Laaser, C.-F., & Rosenschon, A. (2016). *Subventionen in Deutschland bis zum Jahre 2015/2016 – Das Geld sitzt deutlich lockerer* (Kieler Beiträge zur Wirtschaftspolitik, No. 9/2016). Kiel Institute for the World Economy. Available at https://www.ifw-kiel.de/pub/wipo/volumes/wipo_9.pdf.

Leipold, H. (2006). *Kulturvergleichende Institutionenökonomik*. Stutgart: UTB, Lucius & Lucius.

Lorch, A. (2013). *Vom Ordoliberalismus zum substantiellen Liberalismus – Grundlagen einer freiheitlichen Gesellschafts- und Wirtschaftsordnung.* St. Gallen: Universitätsverlag.

Marquardt, A., & Bagus, P. (2017). *Wir schaffen das - alleine!* München: FinanzBuch Verlag.

Mill, J. S. (1869/1999). *On Liberty.* London: Longman, Roberts & Green, 1869; New York: Bartleby.com, 1999. http://www.bartleby.com/130/.

Mussler, W. (2011). EU Economic Governance: The German View. In J.-F. Jamet, W. Mussler, & S. De Corte (Eds.), *EU Economic Governance: The French and German Views* (pp. 43–74). Brussels: Centre for European Studies.

Nedergaard, P., & Snaith, H. (2015). 'As I Drifted on a River I Could Not Control': The Unintended Ordoliberal Consequences of the Eurozone Crisis. *Journal of Common Market Studies, 53*(5), 1094–1109.

New York Times. (2016, July 4). After Brexit, Can Germany Lead Europe Alone? Available at https://www.nytimes.com/2016/07/05/opinion/after-brexit-can-germany-lead-europe-alone.html?mcubz=0.

North, D. C. (1990). *Institutions, Institutional Change and Economic Performance.* Cambridge, UK: Cambridge University Press.

Olson, M. (2000). *Power and Prosperity: Outgrowing Communist and Capitalist Dictatorships.* Oxford: Oxford University Press.

Plickert, P. (2016). *Die VWL auf der Sinnsuche.* Frankfurt/M: Frankfurter Allgmeine Zeitung-Verlag.

Raico, R. (2013, May 10). The European Miracle. *Mises Daily Articles.* Available at https://mises.org/blog/no-country-should-be-bigger.

Röpke, W. (1951/2009). Zu spät und nicht zu spät. Europa als geistige, politische und wirtschaftliche Aufgabe. In W. Röpke (Ed.), *Wilhelm Röpke. Marktwirtschaft ist nicht genug, Gesammelte Aufsätze* (pp. 224–234). Berlin: Manuscriptum.

Röpke, W. (1955). Europäische Investitionsplanung, Das Beispiel der Montanunion. *ORDO—Jahrbuch für die Ordnung von Wirtschaft und Gesellschaft, 7,* 71–102.

Röpke, W. (1959). *International Order and Economic Integration.* Dordrecht, Holland: D. Reidel Publishing Company.

Röpke, W. (1961/1964). Europa in der Welt von heute. In W. Röpke (Ed.), *Wort und Wirkung* (pp. 292–309). Ludwigsburg: Martin Hoch.

Röpke, W. (1966). *Torheiten unserer Zeit, Stellungnahmen zur Gegenwart.* Nürnberg: Glock und Lutz.

Rüstow, A. (1961). *Die Kehrseite des Wirtschaftswunders*. Hamburg: Fuche.

Sally, R. (2016, April 16). *Ludwig Erhard's Social Market Economy—A Liberal, Not a Social Democratic Concept*. Institute of Economic Affairs. Available at https://iea.org.uk/blog/ludwig-erhards-social-market-economy-a-liberal-not-a-social-democratic-concept.

Schüller, A. (2001). Christliches Menschenbild und das Grundproblem der Sozialpolitik. In L. Gerken & J. Starbatty (Eds.), *Schlesien auf dem Weg in die Europäische Union* (pp. 57–65). Oldenbourg: De Gruyter.

Schüller, A. (2002). Soziale Marktwirtschaft. In A. Schüller & H.-G. Krüsselberg (Eds.), *Grundbegriffe der Ordnungstheorie und Politischen Ökonomik* (pp. 156–163). Arbeitsberichte zu Ordnungsfrage der Wirtschaft, No. 7, 5. durchgesehene und ergänzte Auflage. Marburg: Marburger Gesellschaft für Ordnungsfragen der Wirtschaft (MGOW).

Schüller, A. (2011). Das fatale Einheitsdenken in der EU. Lehren aus Selbsttäuschungen und Fehlschlägen, Hayek-Vorlesung. *ORDO—Jahrbuch für die Ordnung von Wirtschaft und Gesellschaft, 62*, 491–515.

Sievert, O. (2003). Vom Keynesianismus zur Angebotspolitik. In Sachverständigenrat zur Begutachtung der gesamtwirtschaftlichen Entwicklung (Eds.), V*ierzig Jahre Sachverständigenrat 1963–2003* (pp. 34–46). Wiesbaden: Statistisches Bundesamt.

Smith, R. (2016, March 30). The liberal Case for 'Leave'. *Adam Smith Institute*. https://www.adamsmith.org/the-liberal-case-for-leave/.

Spiegel-Online. (2017, March 1). *Weißbuch zur Zukunft der EU Fünf Wegweiser, keine Richtung*. http://www.spiegel.de/politik/ausland/analyse-weissbuchder-kommission-zur-zukunft-der-eu-a-1136871.html.

Starbatty, J. (2013). *Tatort Euro. Bürger, schützt das Recht, die Demokratie und euer Vermögen*. Berlin, Wien, and München: Europa Verlag.

Stark, J. (2008). *Monetary, Fiscal, and Financial Stability in Europe*. Speech from November 18, 2008. Available at http://www.ecb.europa.eu/press/key/date/2008/html/sp081118_1.en.html.

Steinmeier, F.-W. (2016, July/August). Germany's New Global Role: Berlin Steps Up. *Foreign Affairs*. Available at https://www.foreignaffairs.com/articles/europe/2016-06-13/germany-s-new-global-role.

Stelzenmüller, C. (2015, July 10). *Germany and the Euro Crisis: Not Just for Austerity's Sake*. Brookings Institution. Available at http://www.brookings.edu/blogs/order-from-chaos/posts/2015/07/10-germany-euro-cr.

Straubhaar, T. (2014, December 2). Die Bundesregierung verachtet ihre eigenen Experten. *Die Welt*. Available at https://www.welt.de/wirtschaft/

article134930867/Die-Bundesregierung-verachtet-ihre-eigenen-Experten.html.

Streit, M. (1995). *Ordnungsökonomik: Versuch einer Standortbestimmung* (Diskussionsbeitrag 04-95). Jena: Max-Planck-Institut zur Erforschung von Wirtschaftssystemen.

Streit, M. E. (2000). *Rationale Wirtschaftspolitik in einem komplexen System* (Diskussionsbeitrag 05-2000). Jena: Max-Planck-Institut zur Erforschung von Wirtschaftssystemen.

Streit, M. E. (2011). *Quo vadis Europa? Beiträge zur europäischen Integrationspolitik.* Baden-Baden: Nomos-Verlag.

Streit, M. E., & Wohlgemuth, M. (1997). *The Market Economy and the State: Hayekian and Ordoliberal Conceptions* (Diskussionsbeitrag 06-97). Jena: Max-Planck-Institut zur Erforschung von Wirtschaftssystemen.

Trebesius, U. (2017, September 6). Europäische Verteilungskämpfe nach der Bundestagswahl? *Tichys Einblick.* Available at https://www.tichyseinblick.de/gastbeitrag/europaeische-verteilungskaempfe-nach-der-bundestagswahl/.

Vanberg, V. (1988). "Ordnungstheorie" as Constitutional Economics—The German Conception of a "Social Market Economy". *ORDO, Jahrbuch für die Ordnung von Wirtschaft und Gesellschaft, 39,* 17–31.

Van Esch, F. A. W. J. (2014). Exploring the Keynesian–Ordoliberal Divide. Flexibility and Convergence in French and German Leaders' Economic Ideas During the Euro-Crisis. *Journal of Contemporary European Studies, 22*(3), 288–302. The European Union After the Crisis

Vaubel, R. (1994). The Political Economy of Centralization and the European Community. *Public Choice, 81,* 151–190.

Vaubel, R. (1996). Constitutional Safeguards Against Centralization in Federal States: An International Cross-Section Analysis. *Constitutional Political Economy, 7,* 79–102.

Vaubel, R. (2009). Constitutional Courts as Promoters of Political Centralization. *European Journal of Law and Economics, 28,* 203–222.

Vaubel, R. (2013). Secession in the European Union. *Economic Affairs, 33*(3), 288–302.

Vaubel, R. (2017, March 30). Wettbewerb der Stadtstaaten. *Wirtschaftliche Freiheit.* Available at http://wirtschaftlichefreiheit.de/wordpress/?p=20615&print=1.

von Mises, L. (1927/2002), *Liberalism: In the Classical Tradition.* Translation of the German edition, 1927 by Ralph Raico 1985. The Foundation for Economic Education. Irvington, NY: The Ludwig von Mises Institute, 2002.

Wehler, H.-U. (2014). *Die Deutschen und der Kapitalismus: Essays zur Geschichte*. München: C.H.Beck.

Weidmann, J. (2013, February 11). *Krisenmanagement und Ordnungspolitik*, Walter-Eucken-Vorlesung. Freiburg. Available at https://www.bundesbank.de/Redaktion/DE/Reden/2013/2013_02_11_weidmann.html.

Welt. (2017, May 11). Schäuble offen für Macrons radikale Euro-Zonen-Pläne. Available at https://www.welt.de/politik/deutschland/article164457086/Schaeuble-offen-fuer-Macrons-radikale-Euro-Zonen-Plaene.html.

Wirtz, B. (2017, March 9). *On This Europe Day, Let's Oppose the EU for the Right Reasons*. Foundation for Economic Education. Available at https://fee.org/articles/on-this-europe-day-lets-oppose-the-eu-for-the-right-reasons/.

Wohlgemuth, M. (2001). The Present Relevance of Ordnungstheorie for the Politics and the Economics of the Social Order. In A. Labrousse & J.-D. Weisz (Eds.), *Institutional Economics in France and Germany: German Ordoliberalism Versus the French Regulation School* (pp. 200–243), Studies in Economic Ethics and Philosophy. Berlin and Heidelberg: Springer.

Wohlgemuth, M. (2013a). The Freiburg School and the Hayekian Challenge. *Review of Austrian Economics, 26*(2), 149–170.

Wohlgemuth, M. (2013b). Introduction: German Neo-liberalism and Its Relevance for Austrian Economics. *Review of Austrian Economics, 26*(2), 105–108.

Wohlgemuth, M. (2017a). *Why Germany 'Kant' Be the UK's Perfect Brexit Ally*. Available at http://www.openeuropeberlin.de/why-germany-kant-be-the-uks-perfect-brexit-ally-by-michael-wohlgemuth/.

Wohlgemuth, M. (2017b). *Von der "immer engeren Union" zum "Europa der Clubs", Schweizer Monat, Sonderthema 33: Ein Europa der Zukunft*. Zürich: SMH Verlag AG.

Wohlgemuth, M. (2017c). Political Union and the Legitimacy Challenge. *European View, 16*(1), 57–65.

Wohlgemuth, M. (2017d, August 21). *Germany Votes: Another Four Years of Merkel—But Which Merkel?* Open Europe Berlin. Available at http://www.openeuropeberlin.de/germany-votes-another-four-years-of-merkel-but-which-merkel-by-michael-wohlgemuth/.

Young, B. (2014). German Ordoliberalism as Agenda Setter for the Euro Crisis: Myth Trumps Reality. *Journal of Contemporary European Studies, 22*(3), 276–287.

5

Changes in Germany's European Policy in the Face of Brexit

Józef Olszyński

5.1 Introduction

Germany's European policy,[1] though invariably geared towards ensuring national interests, has been subject to constant change in terms of objectives and means of action. The most significant and widely noticeable change is the evolution of Germany's European policy away from its Western anchoring (*Westbindung*) and the modesty and sensitivity of the "Bonn Republic" period towards an increasingly resolute expression

[1]The term "European policy" is used here in its narrow sense to denote a country's policy on the European Union and unification processes. Cf. Olszyński, J. (2016), Polityka europejska Niemiec w latach 2012–2013, in Olszyński, J., Bielig, A., and Wandel, J. (ed.), *Niemcy i Polska w drodze do "Europa 2020"*, pp. 43–60.

J. Olszyński (✉)
Collegium of World Economy,
SGH Warsaw School of Economics, Warsaw, Poland
e-mail: jozef.olszynski@sgh.waw.pl

A. M. Kowalski (ed.), *Brexit and the Consequences for International Competitiveness*, https://doi.org/10.1007/978-3-030-03245-6_5

of national interests, coupled with a desire and ambition to play the key leadership role in the EU and be a global superpower under the priorities of today's "Berlin Republic".

Germany's European policy, paralleling the process of European integration, has evolved in step with external geopolitical and regional developments as well as the situation within Germany.

Due to the weakness of other actors in Europe and beyond, the global financial crisis of 2008–2009 and the eurozone crisis that began in 2010 markedly strengthened Germany's position and led it to take a leadership role in the EU "as a necessity". This role was formally decreed in a CDU/CSU-SPD coalition agreement when a new government was formed after parliamentary elections in 2013.

The German leadership of the EU, which steered clear of a hegemony, according to most politicians and experts, took on a cooperative tone and was adapted on an ongoing basis to new, radically difficult conditions and challenges. In addition to problems in the functioning of the eurozone, the Greek crisis, and the conflict in Ukraine, a migration crisis erupted, accompanied by growing Euroscepticism and nationalist and separatist tendencies in many EU member states. All these problems culminated with the decision of the British government, following a referendum, for the UK to leave the European Union, a prospect known as Brexit.

Germany's European policy and the country's leadership of the EU were confronted with a completely new challenge, an unprecedented, significant event in the history of European integration.

This chapter seeks to examine what adaptation reactions have appeared in Germany and what changes have taken place in German European policy in the face of Brexit. The problem has been made particularly relevant by the latest changes in international politics (including the implications of elections in the USA and France) as well as in domestic German politics (the formation of a new coalition government after the 24 September 2017 parliamentary elections).

5.2 Potential Changes (Shifts) in the EU and Germany Post-Brexit

5.2.1 Economic Changes

The challenges posed by Brexit to Germany's European policy are reflected by a set of some basic economic indicators.[2]

The EU's population will shrink by 64.4 million to around 450 million as a result of Brexit. The social profile of EU residents will change slightly, as Britain has a higher-than-average birth rate, high life expectancy, multiculturalism related to high immigration and considerable income stratification.[3]

The EU's economic potential as measured by GDP will be reduced significantly. The UK, with a GDP of EUR 2367 billion, is the second largest EU economy. Its GDP is roughly equivalent to the combined GDPs of the 20 smallest member countries. The EU's GDP will shrink by 13%. The average level of affluence will also decrease, as Britain's GDP per capita is about 8% higher than the EU average.[4]

When it comes to Britain's foreign trade with EU countries, it will certainly see trade-creation and trade-diversion effects opposite to those known from customs union theory. Trade volumes are likely to be reduced, which will result in a loss of jobs in both the EU and the UK. For the UK, the cost of changes in foreign trade conditions is being estimated at 1.3% of GDP in the next 10 years.[5]

---

[2]Multivariate simulations of quantitative changes based on sophisticated econometric models are carried out to help shape government economic policy and build business strategies. Cf. Busch, B. (2017), *Produktions-und Lieferverflechtungen zwischen britischen Branchen, der EU und Deutschland*, "IW-Trends", 2/2017, pp. 61–82.

[3]According to *Deutschland in Zahlen*, Ausgabe 2017, IdW, Koeln.

[4]Ibidem, p. 136.

[5]Cf. Capuano, S. (2017), *Moegliche Konsequenzen des Brexit fuer die Handelsbeziehungen zwischen Grossbritannien und der EU*, "Aktuelle Berichte IAB", Nuernberg 2/2017.

For the European Union, including Germany, the cost levels should be significantly lower due to differences in potential and the possibility of easier shifts in supplies and sales to other markets.

Losses are set to result from the UK's exit from the EU customs union. Non-tariff trade barriers will appear even if Britain's future agreement with the EU provides for duty-free trade. A typical example of non-tariff barriers is the need to confirm the origin of goods, a requirement that can prove to be extremely troublesome and expensive under the current conditions of fragmented production. This in particular applies to the chemical and automotive industries in the case of Britain, and to oil processing, the coking industry, the metal industry and automaking in the case of the EU and Germany.[6] The need to recognise the standards and procedures for admitting each individual product to the market (after Britain's exit from the single market) will be another troublesome and cost-intensive factor.

Even though Britain's involvement in intra-EU trade (42% in exports, 53% in imports) is lower than Germany's, it is of great importance to the UK economy. This in particular applies to the trade of services. Britain's services-dominated economy (79% of the workforce, 80% of GDP) has a surplus of EUR 22 billion in the trade of services and a deficit of around EUR 12 billion in the trade of goods.[7] This means that getting rid of comprehensive regulations governing the provision of services in the single internal market may have a painful effect on Britain.

However, some preliminary assessments of the economic impact of Brexit are for the most part optimistic. Immediately after the Brexit referendum, economic sentiment in the UK slumped, but it quickly rebounded. Share prices before long sprang back to levels higher than before the referendum. The pound depreciated by around 10%, leading to a surge in exports. In all, the UK's 2017 GDP growth was expected to be close to original projections.[8] For the time being, there has

---

[6] Ibid., p. 5 and Busch, B. (2017), op. cit.

[7] Capuano, S. (2017), op. cit., p. 4.

[8] Cf. Mathes, J. (2017), *Der Brexit ist ein Rueckschritt fuer die europaeische Integration*, https://www.iwd.de. Accessed 3 July 2017.

essentially been no flight of capital from London's City financial district. Several banks have announced moves to Dublin, Paris or Frankfurt, but this has not been a massive trend and involves a more distant future.[9]

The most immediately visible economic effect of Brexit for the European Union will be the loss of the UK contribution to the budget. This contribution, taking into account the so-called UK rebate, is GBP 12.9 billion a year. Britain is the third largest contributor to the EU budget, after Germany and France. Its net contribution is around EUR 10 billion, or 7% of the EU budget, making Britain the second largest net contributor to the bloc's coffers.[10]

5.2.2 Non-economic Changes

In formal and institutional terms, Brexit will change how individual EU institutions function and operate. As a result of Brexit, 72 European Parliament seats will be vacated, in addition to 24 seats each on the European Economic and Social Committee and the Committee of the Regions, and there will be 29 votes less in the European Council.

A far more important implication of Brexit is the threat of an imbalance after only two of the EU's "big three" powers remain in the bloc. One concrete change in voting procedures, under the Lisbon Treaty, will be that a group of countries with strict public finance discipline and supportive of free trade (Britain, Germany, the Netherlands, Austria and Finland) will lose a blocking minority (at least 75% of the EU population). The group of Mediterranean countries, with protectionist and interventionist preferences, meanwhile, will see their role grow to about 42%, which may pose a threat to EU economic and trade policies.[11]

The most spectacular shift will take place in the EU's military potential. The British armed forces, which consist of the three

---

[9]Ibid., p. 2.

[10]https://europa.eu/revenue-income-pl. Accessed 15 October 2017.

[11]See Sinn, H. W. (2017), Die Bedeutung des Brexit fuer Deutschland und Europa, *Frankfurter Allgemeine Zeitung*, 16 March 2017.

world-renowned components: the Royal Navy, the British Army and the Royal Air Force (RAF), are now 178,000 strong and the second largest in Europe (and the 28th largest in the world), endowed with a budget of GBP 35 billion (1.9% of GDP), the second biggest worldwide.[12] Britain's armed forces are well equipped and experienced and have a tradition going back centuries. This, combined with the country's special relations with the USA, determines its high military value and position as number two in NATO. In addition, the UK is a nuclear superpower. It is an EU member country that strongly defends the unity of NATO and has always opposed ideas to transfer defence and security policy making to the EU level.

This balance of power within the EU will change after Brexit. It is no coincidence that the first initiatives by Germany and France to deepen European integration have focused on Common Security and Defence Policy (CSDP).[13]

Qualitative changes are set to take place within the EU in less tangible areas such as ideology, law, politics and image.

Certainly, Brexit will deprive the EU of a member state that supports a free market, free international trade and doctrines underlying these trends and processes. A group of states with a preference for statist and protectionist policies, led by France, will see their role grow. Impulses from Anglo-Saxon common law will cease in the EU and a socio-economic model calling for a Europe more strongly geared towards social welfare will encounter fewer objections.

Politically, Brexit will markedly reduce the EU's role in the world. The EU is set to lose a member state that is a permanent member of the UN Security Council, a member of the G7 and G20 groups, and a balancing factor within the bloc as part of the Germany-France-Britain triangle. Such a "fragmentation" of forces may produce negative implications in various aspects of international politics, especially in the longer term. The European Union is set to see its position weakened

[12] *Brytyjskie siły zbrojne*, www.psz.pl. Accessed 16 October 2017.

[13] Szubart, K. (2017), *Unia Europejska "dwóch prędkości"? Niemcy i WPBiO po Brexicie*, BIZ No. 281, www.iz.poznan.pl. Accessed 3 July 2017.

against those of the USA, China and Russia in terms of the system of global powers. Britain itself is also set to lose politically.[14]

All the shifts and negative effects of Brexit are combined with reputation damage for the European integration process. Previously treated as irreversible, this process is now being questioned. A question is being asked whether this model of the European order is still valid. This question is being asked both outside the European Union, in the main centres of world politics, and inside the bloc, where there is a growing lack of trust in EU bodies and criticism of the integration process, either in its entirety or in part. With the looming departure of Britain from the EU, the term "disintegration" has taken on a new, realistic dimension, and it cannot be ruled out that other EU members could be tempted to copy this scenario, resulting in a domino effect.[15]

For Germany, Brexit generally means that its position in the EU will increase in relation to other member states, but it also means increased leadership responsibility for the country.

5.3 Prospects for Germany's European Policy Post-Brexit

5.3.1 First Reactions and Opinions in Germany

The news of the British public supporting Brexit in a referendum was met with an immediate, diverse response, including emotional reactions, in Germany.[16]

---

[14]See Moeller, A. (2016), *Die EU ohne Grossbritannien: politische Folgefragen*, http://www.bpb.de/internazionales/europa/brexit/228804. Accessed 3 July 2017.

[15]Grosse, T. G. (2017), *Ku Europie dwóch prędkości. Strategia Niemiec wobec integracji europejskiej*, CAKJ, Kraków; Lippert, B. (2016), *Die EU zwischen der Integration und "Souveraenitaetsreflexen"*, http://www.bpb.de.internazionales/europa. Accessed 3 July 2017; and Moeller A. (2016), op. cit.

[16]The day after the referendum, leaders from EU institutions (including the president of the European Commission, the chief of the European Council, the head of the European Parliament, and the prime minister of the Netherlands, the country holding the rotating EU presidency at the time) declared that the European Union would continue as a bloc of 27 member states and that Brexit would not be the beginning of the end of the EU, https://www.salon24.pl. Accessed 16 October 2017.

Chancellor Merkel, in her first comment, said that the UK's decision to leave the European Union "is a watershed event for Europe and for the European process of unification and integration", and she appealed for calm and prudence.[17] She also said that "Germany has a special interest and a special responsibility in European unity succeeding".[18] In broad terms, she was speaking on behalf of the entire German government.

In more detailed terms, there were distinct divisions within the federal government in the face of Brexit. The government coalition parties profiled their positions, driven by their traditional views and an ongoing election campaign.

The Social Democrats (SPD) spoke in favour of neutralising the implications of Brexit by radically deepening EU integration in security and migration policies as well as the monetary union. SPD leader Sigmar Gabriel and the former German president of the European Parliament, Martin Schulz, announced a 10-point plan for EU reforms under the heading of "Founding Europe Anew".[19] The plan called for boosting public investment in the EU, increasing democracy (a second chamber of parliament) and bringing greater transparency to EU institutions in terms of procedures used. A document published on 27 June 2016 by the then German Foreign Minister Frank-Walter Steinmeier and French Foreign Minister Jean-Marc Ayrault, entitled "A Strong Europe in an Uncertain World", complemented and added precision to the SPD's position.[20] It argued that interested member states should deepen cooperation: in foreign and security policy (e.g. by jointly planning and conducting military operations, establishing naval forces and increasing the powers of the European prosecution service); in migration and asylum policy (by setting up a common border protection

---

[17]Ibid., p. 1.

[18]Ibid., p. 2. To underline the cooperative nature of Germany's leadership role, she invited the French president and the Italian prime minister to urgent consultations.

[19]According to Frymark, K., and Popławski, K. (2016), *Niemcy wobec Brexitu: powrót sporu o Europę dwóch prędkości*, "Analizy OSW", 29 June 2016, https://www.osw.waw.pl. Accessed 16 October 2017.

[20]Ibid., p. 5.

service and a common system of entry permits, by establishing a European Asylum Agency, and by launching a mechanism for the distribution of refugees); and in the economy (by unifying corporate taxation systems, joint taxation of transnational corporations, and increased convergence in the energy sector, the digital economy and vocational education). The German Social Democrats proceeded from the assumption that deeper cooperation of some countries would lead to the emergence of a two- or multi-speed Europe.

Chancellor Merkel's CDU/CSU, meanwhile, has on the whole been urging EU cohesion and unity. Merkel's general proposals for deeper security and economic cooperation in the face of Brexit have raised no controversy among EU countries. An EU reform should be gradual and acceptable to all 27 member states. In particular, according to the CDU and Merkel, it is necessary to strive for a Europe closer to its citizens (economic reforms should help reduce the gap between those who have benefitted the most from globalisation and those who have lost out) and for increased readiness to take greater responsibility for EU foreign policy (individual member states will be unable to cope with crisis challenges on their own).[21] Merkel has argued that Germany and France bear special responsibility for the EU's success, though the EU27 is strong enough to overcome multifaceted damage done by Brexit.

German politicians, experts and media outlets alike predominately voiced critical and pessimistic views and assessments after the British referendum. It was interpreted as a sign of constructive criticism, a symbol of opposition to leaders, a triumph of populism and a protest against a declining quality of society in Britain. Some highly critical assessments were expressed, including suggestions that would Brexit mark the "beginning of the end of the European Union",[22] that it would harm the German car industry, and that it would represent a major shock to Germany and the EU as a whole. Juergen Matthes of the German Economic Institute (IdW) described Brexit as a clear step backward in European integration, while Prof. Hans-Werner Sinn from

[21]Ibid., p. 4.

[22]wallstreet-online.de.

Munich called Brexit a "medium-sized disaster" (*mittlere Katastrophe*), saying that it is not about the departure of just one of the member states, but of the EU's second largest economy, a UN Security Council member, a nuclear weapon power, and a member state counterbalancing a statist France.[23]

Some German media outlets also published opinions highlighting the positive aspects of Brexit for the EU and Germany. These included the fact that Brexit could be an opportunity for the German city of Frankfurt am Main to become the financial capital of Europe, that the German stock exchange might gain importance, that US businesses would take a greater interest in Germany's stable and predictable economy, that an opportunity would present itself to create a European army, that this might be the last chance to introduce reforms and fend off technocracy in the EU.

5.3.2 Germany's European Strategy in the Context of Brexit

Both best- and worst-case scenarios were considered in the context of Germany's strategy on continued European integration amid efforts to prevent Brexit from breaking up the EU. As German politicians prepared to hold talks to form a new government coalition, prospects ranged from gradual, flexible and non-divisive EU reforms (under a concept advanced by the CDU and Merkel) to attempts to deepen integration by going ahead with the idea of a two- or multi-speed Europe (a concept urged by the German, French and Italian Social Democrats, backed by French President Emmanuel Macron).

Regardless of how the situation was to develop in the course of further negotiations, politicians and game theory experts generally expected Britain to secure a status similar to that of Norway in relations with the EU.

The German strategy began taking shape on the basis of the country's former positions, while also taking into account new, diverse challenges.

[23] Sinn H.-W. (2016), op. cit., and Matthes J. (2017), op. cit.

Chancellor Merkel found herself working under external pressure from the US president as well as the French president and his southern European allies within the eurozone. Domestically, she was forced to reconcile the interests of potential coalition partners during government formation talks.

US President Donald Trump at the start of his presidency made some critical remarks about Germany and indirectly spoke in favour of disintegration trends in Europe.[24] The fact that Europe can no longer count on the USA in security issues to the same extent as in the past, appears to make the prospect of modifications in EU defence policy more likely.[25] Expected further US policy in this area will probably make it difficult for Germany to choose methods of action to preserve the unity of the EU, but it is also likely to lead to a situation in which German leadership methods in this organisation will become more flexible.

The opposite appears to be true of the French president, who, allied with the Social Democrats, is a strong advocate of a multi-speed Europe and pushing for protectionist and interventionist policies to improve the economy and reform the eurozone and to shape economic relations with EU partners, including post-Brexit Britain.

Merkel has found herself in a difficult situation in which she is determined to help President Macron confront Marine Le Pen's increasingly popular National Front party in presidential elections in 2022. This means that Germany is likely to strike a compromise on a multi-speed EU, a scenario inviting a conflict with Eastern European countries.[26] Merkel signalled such a position in a statement at an informal EU summit in Malta on 3 February 2017 when she said that "there will be an EU with different speeds, that not everyone will take part in the same levels of integration".[27]

Merkel is beginning to use this argument as a means of exerting negotiation pressure on countries wary of deeper integration (under a

---

[24]See Grosse, T. G. (2017), op. cit., p. 10.

[25]See Bielecki, J. (2017), *Polska - Niemcy: znikająca wspólnota interesów*, http://www.rp.pl/analizy309219867. Accessed 28 September 2017.

[26]Ibid., p. 2.

[27]Grosse, T. G. (2017), op. cit., p. 7.

compromising approach) and to stop disintegration processes provoked by Britain's exit from the EU.

Meanwhile, the Free Democratic Party (FDP) and the Greens, the CDU's would-be coalition partners, at the time of the government formation talks, voiced opinions that could work in favour of an optimistic scenario in terms of Germany's European policy.

The FDP was clearly opposed to ideas of deepening integration around the eurozone. The party was not only against ideas such as a separate budget and a common eurozone finance minister, but insisted that some elements of eurozone architecture, such as the European Stability Mechanism, should be scrapped.[28]

The Greens were more moderate when it came to both the future shape of the EU and the bloc's eastern policy.

The German strategy covered issues including economic improvement in the eurozone, the migration crisis and an offensive towards closer integration in EU defence policy.[29]

The new German government was likely to follow the main directions of the country's economic policy domestically and within the EU. This also applies to austerity policies and a gradual reduction in the monetary expansion of the European Central Bank (ECB). In terms of closer eurozone integration towards a fiscal union, Merkel's cautious and non-confrontational stand will likely be modified to take into account calls for a common budget, a common eurozone finance minister and common ECB bonds.[30] As a result, Germany will probably support a compromise on partial implementation of fiscal union objectives.

According to the German line of reasoning, economic revival in the eurozone would be promoted—apart from cohesion policy and the so-called Juncker Plan—by new instruments, including deeper

[28]Haszczyński, J. (2017), *Pociąg z Merkel nam nie odjedzie*, http://www.rp.pl/wybory-w-Niemczech. Accessed 21 September 2017.

[29]See Grosse, T. G. (2017), op. cit., pp. 13–20.

[30]The aforementioned economist H. W. Sinn has voiced a view that Germany will be forced to support the fiscal union because this will enable it to hide expected losses generated by the monetary union. See Kozieł, H. (2017), *Niemiecki dryf ku unii fiskalnej*, http://www.rp.pl/gospodarka309209909. Accessed 21 September 2017.

integration in energy policy (regulatory changes towards reducing carbon emissions would encourage new investment projects) and a common EU defence policy. This last element of strategy followed up on an earlier German government position and is also designed to be a means of preventing further decomposition of the EU in the wake of Brexit. After Britain's exit from the EU, a common European defence policy would become a realistic project.

Guidelines in this area were adopted at a European Council meeting in November 2016 and confirmed at an EU summit in December 2016. They provided for[31]: the establishment of a non-military mission headquarters; the launch of battle groups (numbering around 1000 troops from different member states) and a Eurocorps (7000 officers and 60,000 troops); the establishment of a European Defence Fund (to provide credit for arms purchases and military research)[32]; and the introduction of the so-called European Defence Semester (annual reviews of military capabilities and defence potential).

The plan to enhance the Common Security and Defence Policy (CSDP) was praised for its breakthrough role in this area, while being criticised for its insufficient scope and the risk of generating further divisions within the EU into richer and less affluent countries. The latter countries could have problems taking advantage of CSDP and continuing offset programmes in their own industries.

The German government supported the idea of developing the CSDP, describing it as a key pillar of German security and a platform for articulating Germany's strategic interests. It said that the CSDP "may be the last attempt to achieve the global ambitions of the EU in the current institutional form".[33]

---

[31]Grosse, T. G. (2017), op. cit., p. 17.

[32]The EDF will start operating under the EU's new financial framework after 2021 and will be endowed with an annual research budget of around EUR 500 million plus an additional EUR 5 million for the purchase of weapons; EUR 90 million has been set aside for the pilot programme in 2017–2019. Ibid., p. 14.

[33]Szubart, K. (2017), Unia Europejska "dwóch prędkości"? Niemcy i WPBiO po Brexicie, "Biuletyn Instytutu Zachodniego No. 281", www.iz.poznan.pl. Accessed 21 September 2017.

Although Brexit was in part evidently provoked by the migration crisis in Europe and the UK, Britain's impending departure has only indirectly influenced the new migration strategy of the EU and Germany. The original German stance on migration and the refugee crisis, based on a gesture by Chancellor Merkel to open the border in the summer of 2015, was criticised on many sides and consequently modified as migration assumed unexpected proportions. Internally, regulations were introduced in Germany to stem the influx of immigrants, coupled with a faster deportation procedure for those who have committed criminal offences. Externally, Chancellor Merkel, on the EU's behalf, in the spring of 2016 brokered a deal with Turkey that stopped the massive inflow of migrants and refugees via the Turkish–Greek route and initiated work on a new migration and asylum policy. Germany's strategic thinking evolved from full openness and liberalism to restrictions and controls to stem the tide of migrants.

Chancellor Merkel laid down a set of guidelines for shaping future EU-British relations[34]:

- decisions should be made jointly by all 27 member countries;
- the German government will pay special attention to the interests of German citizens and enterprises;
- Britain, after leaving the EU, should be given less favourable terms of developing business relations with the EU than those it enjoyed in its role as a member state;
- the four fundamental EU freedoms must be guaranteed in market access negotiations: movement of persons, services, capital and goods.

Experts from the German Economic Institute defined three critical negotiation areas for detailed discussions. In their opinion, access to the EU single internal market, the freedom of movement of persons and UK payments to the EU budget remained to be discussed. Depending on what kind of negotiation strategy were to be used: a hard Brexit (uncompromising approach) or a soft exit (with the EU and the UK

[34]Frymark, K., and Popławski, K. (2016), op. cit., p. 4.

tending to compromise), different options of negotiation outcomes appeared to be possible: from a relationship based on WTO rules (both sides uncompromising) to the most-expected and apparently most realistic option involving "Norway-plus status" (tendency to compromise), in which Britain would continue contributing to the EU budget in exchange for access to the single internal market and some limited freedom of the movement of persons.[35]

Future German government strategy on Brexit will likely be defined as ranging between "deterrence" of possible future imitation to ensuring minimum possible damage on both sides. Chancellor Merkel has been quoted as recommending staying "calm, composed and determined, while taking into account the need for a partnership between the new European Union and Britain".

5.3.3 New Character of German Leadership?

Brexit will significantly change Germany's position in the European Union in many areas and add to disproportions in geopolitical potential with regard to France, the bloc's second-largest member. Germany's share of the EU population and the bloc's economic and military potential will increase markedly, and the country's role in culture, science and technology will grow as well. The German socio-economic model (*soziale Marktwirtschaft*) and the country's legal system will become even more dominant in the European Union, with no counterbalance from the Anglo-Saxon model. The EU will be increasingly identified with Germany.

Overall, in quantitative terms, Germany's leadership of the European Union stands to be strengthened. At this point, it is impossible to predict whether there will also be qualitative changes, i.e. changes in the nature of leadership, and how profound they may be. This depends on multiple factors inside Germany and externally, both within the EU and globally.

---

[35] *Die Loesung heisst Norwegen*, https://www.iwd.de/artikel/322535. Accessed 3 July 2017.

Historically, the process of European integration, which in its initial stages was essentially "an idea for systematically limiting Germany's role in Europe", with time became a tool for increasing the country's dominance.[36] It led to Germany taking over a leadership role, first out of necessity (2010–2012) and then (from 2013 onwards) in a conscious and purposeful process, amid declarations of assuming responsibility for the success of the European project. Eventually, a situation developed where Germany felt compelled to use its growing power to stop the EU from breaking up in the face of Brexit. Politicians and experts agree that German leadership in the EU is not hegemonic and that Brexit does not change that.

The German government, in its own interest amid efforts to prevent attempts to revive the so-called German question,[37] is trying to transform its negatively perceived semi-hegemonic position into an EU leadership role defined in terms of cooperative leadership. Former German Foreign Minister Joschka Fischer has said: "Europe will never work along the lines of a hegemony. If anyone seeks to be hegemonic, then everyone, or almost everyone, else unites against them. So the EU will never pan out as a German project".[38]

Germany's leadership role every now and then causes an increasingly large group of EU countries to feel frustrated and dissatisfied. The Slovaks, Czechs and Hungarians have joined the Spaniards, Italians and Greeks in "feeling harmed and angered" by Chancellor Merkel. There are well-known fears of German dominance in Poland, and even France sometimes feels humiliated.[39]

Demands to strengthen German leadership in Europe, including in the context of Brexit, have coexisted with proposals to limit its hegemonic position. At the same time, calls for a decisive German stance in Brexit negotiations have been accompanied by appeals

---

[36]Grosse, T. G. (2017), op. cit., p. 10.

[37]See Kędzierski, M. (2016), *Europejskie Niemcy w niemieckiej Europie*, www.psz.pl. Accessed 3 July 2017.

[38]Fischer, J. (2015), Fatale Entscheidung fuer ein deutches Europa, *Sueddeutsche Zeitung*, 26 July 2015.

[39]Kędzierski M. (2016), op. cit., p. 6.

for more empathy, magnanimity and benevolence on the part of Germany.[40]

But of particular importance in the context of the future shape of the EU is the imperative of cooperation between all member states. Germany is acting as a team leader. It will be working for the benefit of the whole team in this role as long as the right conditions are created for that.

References

Bielecki, J. (2017). *Polska - Niemcy: znikająca wspólnota interesów*. http://www.rp.pl/analizy309219867. Accessed September 28, 2017.

Brytyjskie siły zbrojne, wg www.psz.pl. Accessed October 16, 2017.

Busch, B. (2017). *Produktions-und Lieferverflechtungen zwischen britischen Branchen, der EU und Deutschland* (IW-Trends, 2/2017), pp. 61–82.

Capuano S. (2017), *Moegliche Konsequenzen des Brexit fuer die Handelsbeziehungen zwischen Grossbritannien und der EU* (Aktuelle Berichte IAB, Nuernberg 2/2017).

Deutschland in Zahlen, Ausgabe 2017. Koeln: IdW.

Die Loesung heisst Norwegen. https://www.iwd.de/artikel/322535. Accessed July 3, 2017.

Fischer, J. (2015, July 26). Fatale Entscheidung fuer ein deutches Europa. *Sueddeutsche Zeitung*.

Frymark, K., & Popławski. K. (2016, June 29), *Niemcy wobec Brexitu: powrót sporu o Europę dwóch prędkości*. Analizy OSW. https://www.osw.waw.pl. Accessed October 16, 2017.

Grosse, T. (2017). *Ku Europie dwóch prędkości. Strategia Niemiec wobec integracji europejskiej*. Kraków: CAKJ.

Haszczyński, J. (2017). *Pociąg z Merkel nam nie odjedzie*. http://www.rp.pl/wybory-w-Niemczech. Accessed September 21, 2017.

[40]This involves, for example, suggestions that the German government should avoid playing a role as a teacher and fastidious judge who demands that rules are dutifully followed and punishes any transgressions; and that instead it should create conditions for an enhanced atmosphere of trust. Cf. Maul, H. W. (2015), Uebermaechtig und verwundbar. Deutschlands Rolle in Europa, *Internationale Politik*, No. 5, pp. 132–136.

http://www.bpb.de.internazionales/europa. Accessed July 3, 2017.
https://europa.eu/revenue-income-pl. Accessed October 15, 2017.
https://www.salon24.pl. Accessed October 16, 2017.
Kędzierski, M. (2016). *Europejskie Niemcy w niemieckiej Europie*. www.psz.pl. Accessed July 3, 2017.
Kozieł, H. (2017). *Niemiecki dryf ku unii fiskalnej*. http://www.rp.pl/gospodarka309209909. Accessed September 21, 2017.
Lippert, B. (2016). *Die EU zwischen der Integration und "Souveraenitaetsreflexen"*. http://www.bpb.de.internazionales/europa. Accessed July 3, 2017.
Mathes J. (2017). *Der Brexit ist ein Rueckschritt fuer die europaeische Integration*. https://www.iwd.de. Accessed July 3, 2017.
Maul, H. W. (2015). Uebermaechtig und verwundbar. Deuschlands Rolle in Europa. *Internationale Politik* (5), 132–136.
Moeller, A. (2016). *Die EU ohne Grossbritannien: politische Folgefragen*. http://www.bpb.de/internazionales/europa/brexit/228804. Accessed July 3, 2017.
Olszyński, J. (2016). Polityka europejska Niemiec w latach 2012–2013. In J. Olszyński, A. Bielig, & J. Wandel (Eds.), *Niemcy i Polska w drodze do "Europa 2020"* (pp. 43–60). Warsaw: Szkoła Główna Handlowa w Warszawie – Oficyna Wydawnicza.
Sinn, H. W. (2017, March 16). Die Bedeutung des Brexit fuer Deutschland und Europa. *Frankfurter Allgemeine Zeitung*.
Szubart, K. (2017). *Unia Europejska "dwóch prędkości"? Niemcy i WPBiO po Brexicie* (BIZ No. 281). www.iz.poznan.pl. Accessed July 3, 2017.

6

Bilateral Trade and Investment Between Britain and Germany Ahead of the UK's Impending Departure from the European Union

Andreas Bielig

6.1 Introduction

It is not an exaggeration to say that the Brexit referendum decision on 23 June 2016 caught international markets by surprise. In Germany, only diehard pessimists were advancing the worst-case scenario of the UK leaving the EU, but that was exactly what happened. With the required minimum of 51.9% of the vote, 17,410,742 British citizens voted for the exit option, while only 16,141,241 voted "Remain". Although the UK had long played a distinctive role in the EU, most economic decision-makers in Germany were surprised by the new signals of dissatisfaction with the Union coming from Britain. In Germany, the UK was seen as a key business partner and a stabilising element of the political order in Europe. Even though they differ substantially on future political developments in Europe, economically the

A. Bielig (✉)
Collegium of World Economy,
SGH Warsaw School of Economics, Warsaw, Poland
e-mail: abielig@sgh.waw.pl

A. M. Kowalski (ed.), *Brexit and the Consequences for International Competitiveness*, https://doi.org/10.1007/978-3-030-03245-6_6

two countries will remain natural partners for cooperation in the future. The track record of bilateral economic cooperation between Germany and the UK is impressive in both quantitative and qualitative terms. With Zero Hour fast approaching, the rules of this partnership must be readjusted and economic decisions and projects must be adapted to the new policy regime post-Brexit. This holds for many areas of economic activity of which two are analysed here: trade development and foreign investment. This chapter analyses trade between Germany and the UK from 2008 to 2016 using descriptive statistics, and it also traces the development of bilateral direct investment, providing an insight into the potential implications of Brexit for two-way exchange.

6.2 Development of Trade Between Germany and UK from 2008 to 2016

Trade between Germany and the UK has developed successfully in the last few decades. In the 1990–2016 period, German exports to Britain grew by 204.8%, from EUR 28.26 billion in 1990 to EUR 86.15 billion in 2016, with average annual exports at EUR 52.83 billion. After a period of stagnation in the early 1990s, exports increased each year until the start of an economic crisis in 2007, and then rose again from 2010 onwards. In 2016, when 51.9% of the British people voted in favour of their country leaving the EU, exports decreased by 3.2%, as shown in Fig. 6.1.[1] In 2016, the UK was ranked third among Germany's largest export partners,[2] so potential trade barriers in the future would pose an immense risk. Also, 12.2% of Germany's EU exports were to the UK (2nd place after France).[3] The fastest growth in exports was recorded in 1997, at 17.7%, whereas the greatest decrease took place in 2009, at 17.0%. Exports grew at a respectable annual rate of 4.7% in 1990–2016, so bilateral trade between Germany and UK can be described as a success story for both partners.

[1]Destatis (2017a).

[2]Destatis (2017b).

[3]Destatis (2017c).

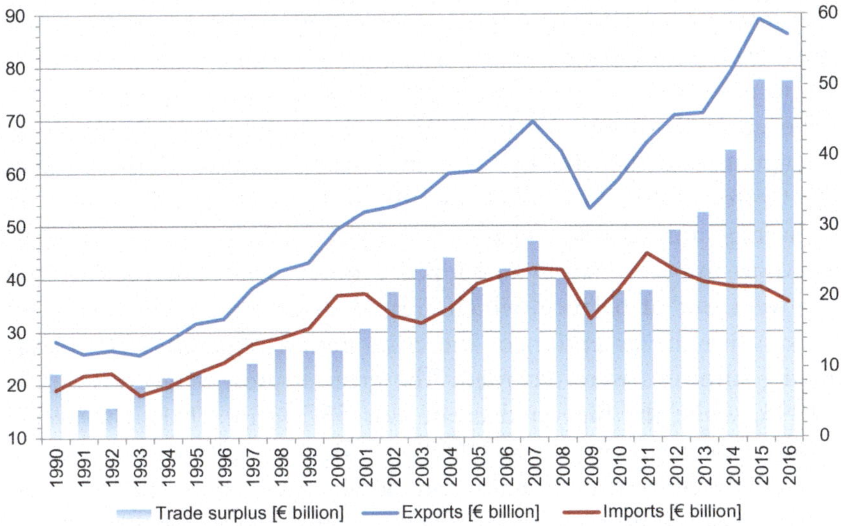

Fig. 6.1 Bilateral trade between Germany and UK, 1990–2016, exports and imports [EUR billion] left axis, trade surplus [EUR billion] right axis (*Source* Own elaboration on basis of Destatis [2017a])

The same conclusions, with some minor modifications, hold true for imports. The UK is not among Germany's top 10 import partners, being ranked in 11th position in 2016,[4] which nevertheless means that this direction of trade flows plays a substantial role. The UK was in eighth place among Germany's largest import partners, with 6.5% of total imported goods.[5] Imports increased by 86.6% from EUR 19.1 billion in 1990 to EUR 35.7 billion in 2016. Even if the growth trend in imports was not as stable as in the case of exports (with a remarkable decrease in 2002 and 2003), imports in general—except in crisis years—followed an expansion trend, with annual growth rates of 3.0%. The greatest growth of 18.0% was recorded in 2011, whereas the largest decrease was in 2009 at 22.1%. Average imports in 1990–2016 were EUR 32.63 billion. Since 2012, the import growth path has appeared

[4]Destatis (2017b).

[5]Destatis (2017c).

to be interrupted and uncoupled from export dynamics, with average annual growth rates negative at 4.4%. These shrinking German imports seem to point to a declining competitiveness of British exporters.

The growing divergence between export and import dynamics led to a remarkable increase in Germany's trade surplus. It increased by 452.1% from EUR 9.14 billion in 1990 to EUR 50.45 billion in 2016. The average annual trade surplus was EUR 20.21 billion. Until 1996, the export-import relation developed symmetrically but then increasingly went out of balance, which created space for growing trade surpluses. With the exception of the crisis period of 2008–2010 rapidly growing surpluses were recorded. The greatest growth rate was 76.9% in 1993, while the biggest decrease was 55.3% in 1991. The average growth of the trade surplus was 9.6% p.a. But the 0.3% drop in the surplus in 2016 hints at potential future developments after the full implementation of the Brexit agenda, which remains unclear for now.

By far, *machinery and transport equipment* is the dominant category of products traded between Germany and the UK at the one-digit level of the Standard International Trade Classification (SITC). A new SITC classification by the Statistical Office in Wiesbaden provides data from 2008 onwards, whereas previous data use an older national classification system that is not fully compatible with the current SITC standards. Exports of machinery increased by 39.1% from EUR 33.96 billion in 2008 to EUR 47.25 billion in 2016,[6] as shown in Fig. 6.2. With the exception of 2009, export volumes increased steadily, with the highest growth rate in 2015 at 16.8%. The largest relative drop in exports was recorded in 2009, at 18.6%. Between 2008 and 2016 the average volume was EUR 37.45 billion, while absolute volumes expanded at a rate of 4.8% a year on average. At the same time, the share of machinery in total UK exports grew from 52.9% to 54.8%, with the average share at 53.0%. The greatest growth of export shares was recorded in 2009 at 4.0%, yet the average annual growth, due to considerable fluctuations, was only 0.5%.

[6]Destatis (2017d).

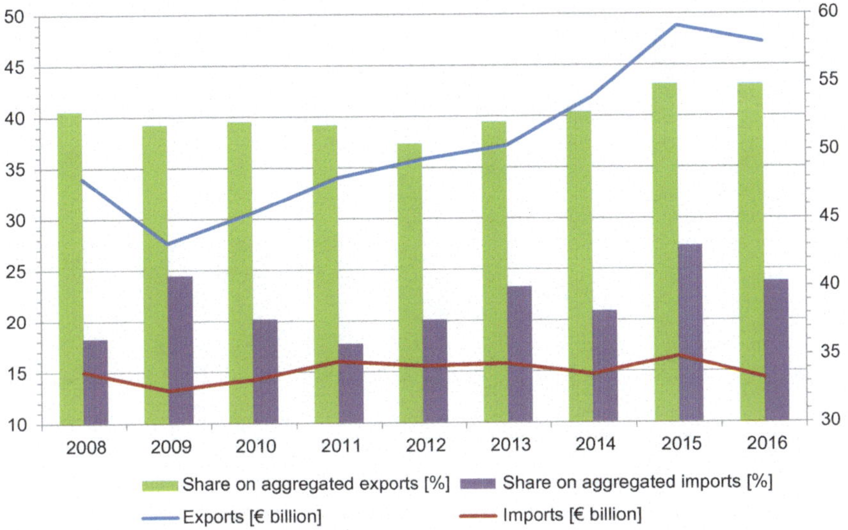

Fig. 6.2 Bilateral trade between Germany and the UK, 2008–2016, machinery and transport equipment (SITC-7), left axis exports and imports [EUR billion], right axis shares of aggregated exports and imports [%] (*Source* Own elaboration on basis of Destatis [2017d])

German machinery imports from the UK showed a stagnation during the same period. They declined by 4.4% from EUR 15.08 billion in 2008 to EUR 14.41 billion in 2016. Their development followed a path nearly symmetrical to that of the average level of imports, which stood at EUR 15.07 billion and showed a negative average annual growth rate of 0.1%. Despite the shrinking absolute volumes, the share of machinery in total German imports from the UK increased from 36.2% in 2008 to 40.4% in 2016, with an average share of 39.0%. The figures show that machinery was less prominent in German imports than in exports in terms of absolute shares, but it nonetheless represented the largest product group. The greatest growth was recorded in 2009 at 12.8%, whereas the biggest decline was recorded in 2010, at 7.9%. The average import share of machinery grew by 1.7% p.a.

Chemicals and related products constituted the second largest product group in terms of Germany's export volumes to the UK in 2016. Exports of chemicals increased by 55.1% from EUR 8.26 billion in

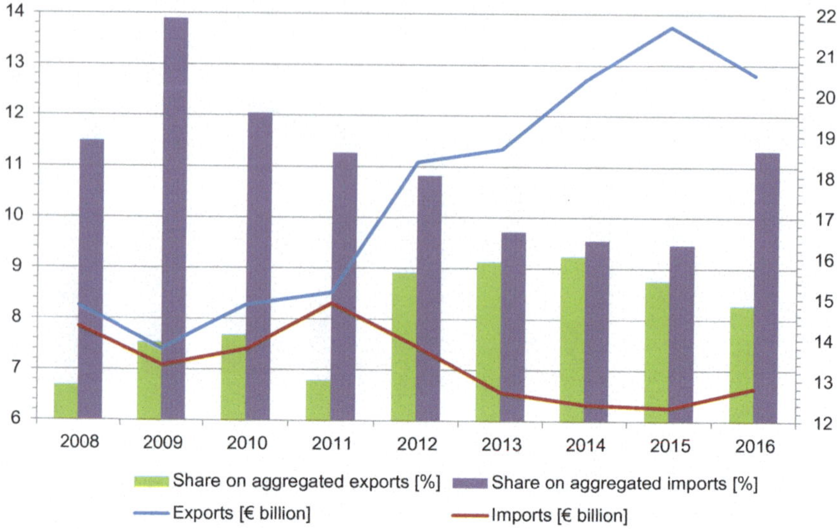

Fig. 6.3 Bilateral trade between Germany and the UK, 2008–2016, chemicals and related products, n.e.s. (SITC-5), left axis exports and imports [EUR billion], right axis shares of aggregated exports and imports [%] (*Source* Own elaboration on basis of Destatis [2017e])

2008 to EUR 12.80 billion in 2016, as shown in Fig. 6.3.[7] Reduced volumes were recorded in 2009. The highest annual growth rate, 30.1%, was recorded in 2012, while the steepest slide was in 2009, at 10.1%. On average, chemicals exports totalled EUR 10.47 billion, with an average annual growth rate of 6.3%. Because of this strong dynamics, which outstripped the average growth of overall German exports to the UK, the share of chemicals in overall exports increased from 12.9% in 2008 to 14.9% in 2016, with an average share of 14.6%. Still, the export performance of chemicals was not constantly higher than average: diminishing shares were recorded on three occasions, with the greatest slide, by 7.9%, in 2011. Curiously, the greatest increase was recorded a year later, at 20.4%. On average, the export share of chemicals grew by 2.1% p.a.

[7]Destatis (2017e).

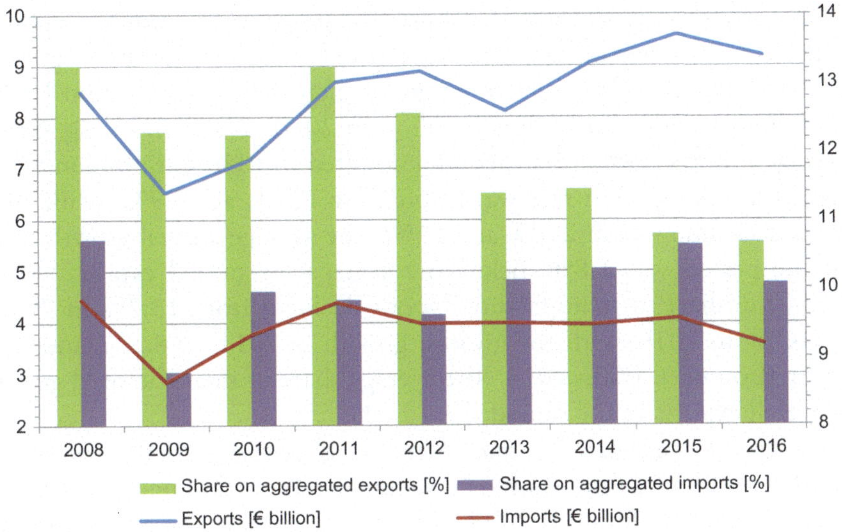

Fig. 6.4 Bilateral trade between Germany and the UK, 2008–2016, manufactured goods classified chiefly by material (SITC-6), left axis exports and imports [EUR billion], right axis shares of aggregated exports and imports [%] (*Source* Own elaboration on basis of Destatis [2017f])

Meanwhile, German imports of chemicals from the UK showed some minor declines in absolute figures. They declined by 15.3% from EUR 7.86 billion in 2008 to EUR 6.66 billion in 2016. After a temporary recovery in 2011, with a peak in import volumes at EUR 8.32 billion, an overall downward trend set in. Average imports amounted to EUR 7.11 billion. The greatest dynamics, at 12.2%, was recorded in 2011, whereas the greatest decline was in 2013, at 12.1%. The average annual growth rate was negative at 1.7%. The import performance of chemicals was worse than that of overall German imports from the UK, as shown in Fig. 6.1. The import share of chemicals decreased from 18.9% in 2008 to 18.6% in 2016. The greatest increase was recorded in 2009, at 15.9%, while the biggest drop was in 2009, at 10.6%. Despite these negative developments, the average share of chemicals grew by 0.2% p.a. The average share of chemicals in overall imports was 18.3%.

Manufactured goods form the third largest group of German exports to the UK. In terms of size, they represented 19.5% of the largest group in 2016. Exports of manufactured goods increased by 8.1% from EUR 8.5 billion in 2008 to EUR 9.2 billion in 2016, as shown in Fig. 6.4.[8] Average exports during that period amounted to EUR 8.42 billion. The highest annual growth rate was recorded in 2011, at 20.9%, while the steepest decline was in 2009, at 23.1%. The average annual growth rate for the period was 1.8%, below the average for overall exports. As a result, the share of manufactured goods declined from 13.2% in 2008 to 10.7% in 2016, with the greatest growth in 2011, at 8.2%, and the largest drop of 9.3% in 2013. The average share of manufactured goods in overall exports in the period was 12.0%.

Meanwhile, imports of manufactured goods exhibited a negative trend in both absolute and relative terms. They declined by 19.4% from EUR 4.46 billion in 2008 to EUR 3.6 billion. The average value of imports in that period was EUR 3.9 billion. After 2012, imports showed a stagnation trend with nearly no volatility compared with the prior period, with an annual growth rate of 32.5% in 2010 and a 36.2% drop in imports in 2009. The average annual growth for the whole period was negative at 0.7%. Also in relation to total imports, the share of manufactured goods declined: from 10.7% to 10.1%. Manufactured goods achieved the greatest gain in terms of their share of overall German imports in 2010, with 13.4%, while the biggest decline occurred in 2009, at 18.9%. In absolute terms, the average annual growth in the share of manufactured goods was negative at 0.4%.

Miscellaneous manufactured articles rank number four among German exports. In quantitative terms, however, the role of these goods is limited. In 2016, they represented 16.9% of the largest export category in 2016. Nevertheless, manufactured goods exports grew by 13.2% from EUR 7.05 billion in 2008 to EUR 7.98 billion in 2016, as shown in Fig. 6.5.[9] The development of exports was characterised by a stagnation trend. Only in 2014 and 2015 were the annual growth rates near the double digits. The largest decrease was recorded in 2009, at 15.0%.

[8] Destatis (2017f).

[9] Destatis (2017g).

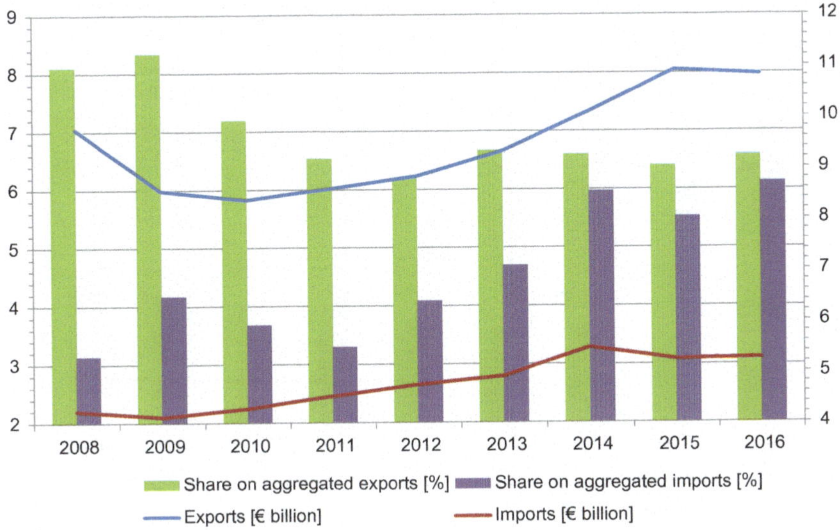

Fig. 6.5 Bilateral trade between Germany and the UK, 2008–2016, miscellaneous manufactured articles (SITC-8), left axis exports and imports [EUR billion], right axis shares of aggregated exports and imports [%] (*Source* Own elaboration on basis of Destatis [2017g])

The average annual growth of manufactured goods exports was nonetheless positive at 1.9%. This limited dynamics led to a 15.7% reduction in the proportion of manufactured goods in total German exports to the UK, from 11.0% in 2008 to 9.7% in 2016. The biggest drop, at 11.6%, occurred in 2010, while the fastest growth was in 2013 at 6.2%, with a negative average annual growth rate of 2.0%. The average share in the period was 9.7%.

Like exports, imports of manufactured goods increased between 2008 and 2016. They rose by 41.0% from EUR 2.21 billion to EUR 3.12 billion. Average imports in that period were EUR 2.66 billion, with a peak of EUR 3.29 billion recorded in 2014. The most dynamic growth was also recorded in 2014, at 17.7%, while the biggest drop, at 6.1%, occurred a year later. Annual imports grew by 4.6% on average. This dynamics was sufficient for "miscellaneous manufactured articles" to increase their share of total German imports from the UK from 5.3% in 2008 to 8.7% in

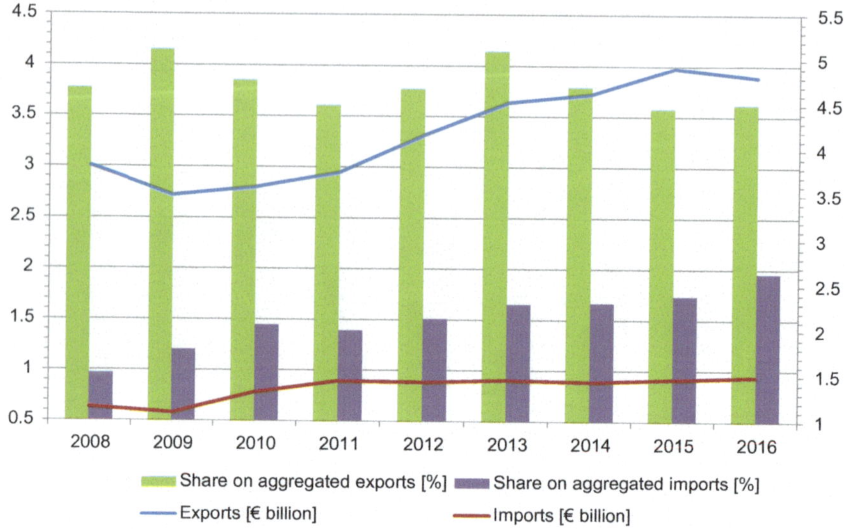

Fig. 6.6 Bilateral trade between Germany and the UK, 2008–2016, food and live animals (SITC-0), left axis exports and imports [EUR billion], right axis shares of aggregated exports and imports [%] (*Source* Own elaboration on basis of Destatis [2017h])

2016. The average share in that period was EUR 6.89 billion. The greatest growth in the share was recorded in 2009, at 22.3%, while the biggest drop took place a year later, at 8.7%. The average share of manufactured goods in overall imports was 7.1%.

Food and live animals are No. 5 among German exports to the UK, according to volumes traded in 2016. They represent 8.2% of the largest export group and are of minor importance to the development of overall trade between the two economies. Exports of food and live animals increased by 29.3% from EUR 3.01 billion in 2008 to EUR 3.89 billion in 2016, as shown in Fig. 6.6.[10] The greatest increase in such exports was recorded in 2012, at 12.5%, while the largest decrease was seen in 2009, at 9.5%, with average annual growth at 3.5%. Average exports were roughly EUR 3.34 billion in the studied period. However,

[10]Destatis (2017h).

the dynamics was insufficient for this group of goods to maintain their role in overall German exports to the UK. The share of food and live animals in total German exports to the UK decreased from 4.7% in 2008 to 4.5% to 2016. The greatest annual growth of 9.1% was observed in 2008, while the greatest slump was recorded in 2014, at 7.9%. The average growth rate was negative at 0.3%.

Imports of food and live animals increased by 48.1% between 2008 and 2016, from EUR 0.64 billion to EUR 0.94 billion. The greatest annual growth was in 2010, at 34.9%, while the steepest decline was reported in 2008, at 8.8%. The average annual growth rate was 5.7% and the average import volume in the period was EUR 0.83 billion. The substantial import dynamics led to food and live animals increasing their share of overall German imports to the UK from 1.5% to 2.6%. The average share in the period was 2.1%, with a growth peak of 17.1% in 2009 and a low of 3.0% in 2011. The import share of this group of goods grew at an average annual rate of 7.3%.

Three other German product groups saw their export volumes increase to the UK: "commodities and transactions not classified elsewhere in the SITC" (up by 193.9% as shown in Appendix, Fig. 6.12); "crude materials, inedible, except fuels" (up by 25.5% as shown in Appendix, Fig. 6.13); and beverages and tobacco (up by 17.2% as shown in Appendix, Fig. 6.15). Meanwhile, two product groups saw downward trends in export volumes. Those were mineral fuels, lubricants and related materials (see Appendix, Fig. 6.14); and animal and vegetable oils, fats and waxes (see Appendix, Fig. 6.16). Only one product group, commodities and transactions not classified elsewhere in the SITC, increased its share of total exports to the UK. On the import side, traded volumes increased in only two groups: commodities and transactions not classified elsewhere in the SITC (up by 118.5%), and beverages and tobacco (up by 44.7%). Four of the five groups increased their shares of total imports, the exception being mineral fuels, lubricants and related materials. However, the largest of these product groups, commodities and transactions not classified elsewhere in the SITC, represented just 6.9% of the top group in terms of overall German exports, so the quantitative importance of these five categories is limited. With the exception of mineral fuels, lubricants and related materials, all the other product groups recorded trade surpluses, which means that German exporters are strongly dependent on their British partners. Brexit could affect both exports and imports in diverse ways.

6.3 German Direct Investment in UK from 2010 to 2014

Another pillar of German–British economic relations is bilateral investment. Like with bilateral trade, direct investment in the last two-and-a-half decades can be described as a success story. German direct investment in the UK increased by an impressive 1626.6%, from EUR 7.7 billion in 1989 to EUR 128.57 billion in 2014, as shown in Fig. 6.7.[11] The average investment in the period was EUR 63.93 billion. The investment growth rates were high until the start of the financial crisis in 2008, with the three-year moving average ranging from 6.8 to 30.0% p.a. After 2008, the growth rates were in the single digits, except for a slight decline in 2009. This marked a substantial deceleration, but the dynamics nevertheless remained remarkable, with the three-year moving average at 6.3% in 2014. The peak of investment growth was recorded in 1995, at 35.7%. The lowest point, meanwhile, was in 2008, when a 16.8% drop took place—though this last figure may be a statistical outlier rather than an outcome of the economic crisis. The average annual growth of German investment in the studied period was 12.8%, chiefly due to its high dynamics in the 1990s. However, the period of extraordinary investment dynamics for German investors in the UK has ended.

The development of British investment in Germany also adds up to a positive picture, though at a significantly lower level. In 2014, UK direct investment in Germany represented only 23.1% of German investment in the UK. Between 1989 and 2014, British investment in Germany increased by 385.6%, from EUR 6.1 billion to EUR 29.7 billion (Fig. 6.8).

[11]Bundesbank (2017a).

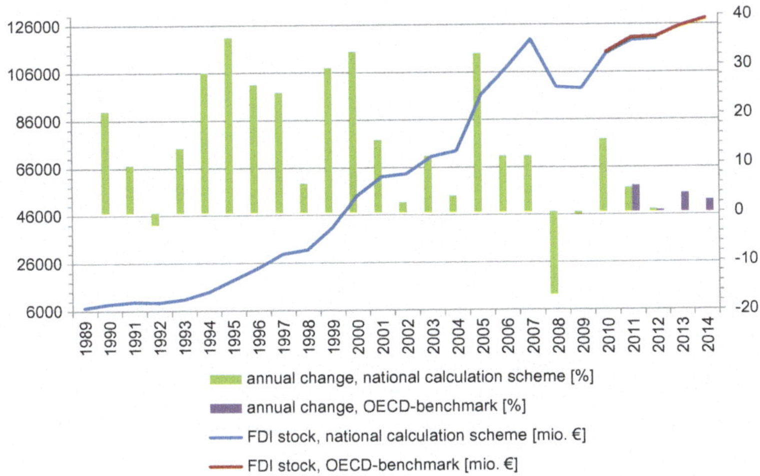

Fig. 6.7 German direct investment in the UK, 1989–2014, stocks, left axis [EUR million], annual change, right axis [%] (*Source* Own elaboration on basis of Bundesbank [2017a])

In 2010, German investors were involved in 1277 enterprises in the UK; by 2014 they had increased their involvement by 7.5% to 1373 enterprises, as shown in Fig. 6.9.[12] The expansion trend showed a 3.6% uptick in 2011; otherwise, growth was limited. The average number of German-owned enterprises in the UK in the studied period was 1333, and the average annual growth rate was 1.85%, which means that the expansion process was steady but relatively slow. At the same time, average investment per enterprise increased by 11.5% from EUR 19.45 million in 2010 to EUR 21.67 million in 2014. Average investment throughout the period was EUR 19.44 million.

Employment at companies run by German investors in the UK increased by 14.5%, from 207,000 in 2010 to 237,000 in 2014, as shown in Fig. 6.10,[13] with the annual growth rate peaking at 5.3% in 2011.

[12]Bundesbank (2017c).

[13]Bundesbank (2017d).

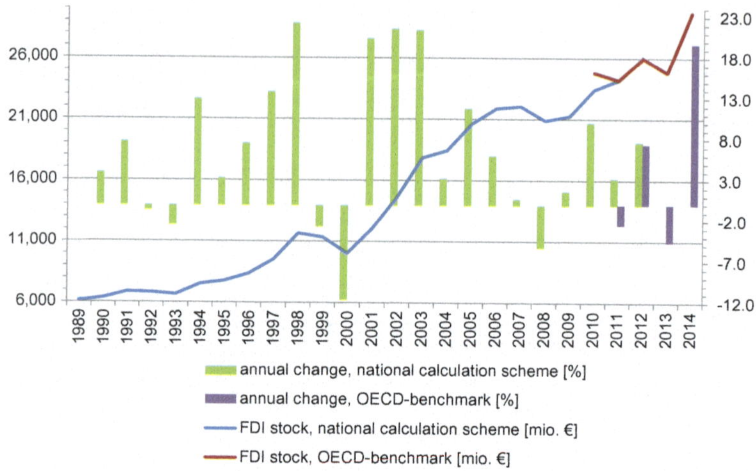

Fig. 6.8 UK direct investment in Germany, 1989–2014, stock, left axis [EUR million], annual change, right axis [%] (*Source* Own elaboration on basis of Bundesbank [2017b])

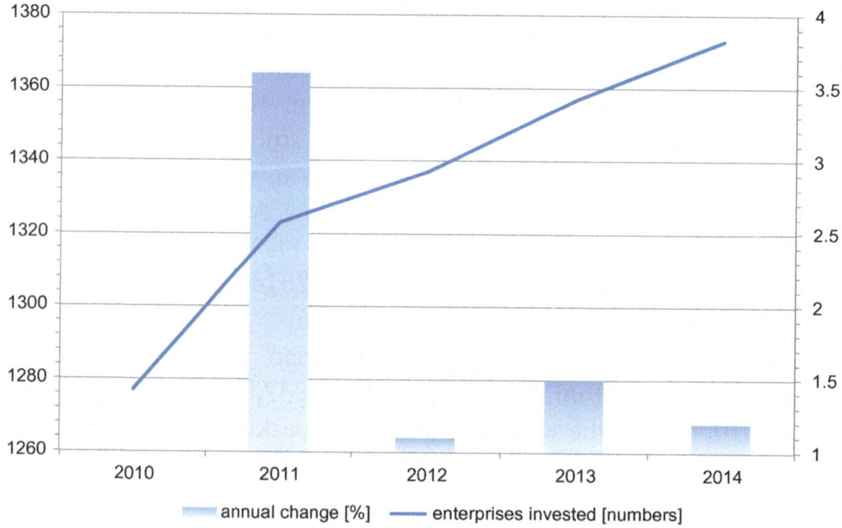

Fig. 6.9 German direct investment in the UK, 2010–2014, enterprises invested left axis [number], annual change [%] (*Source* Own elaboration on basis of Bundesbank [2017c])

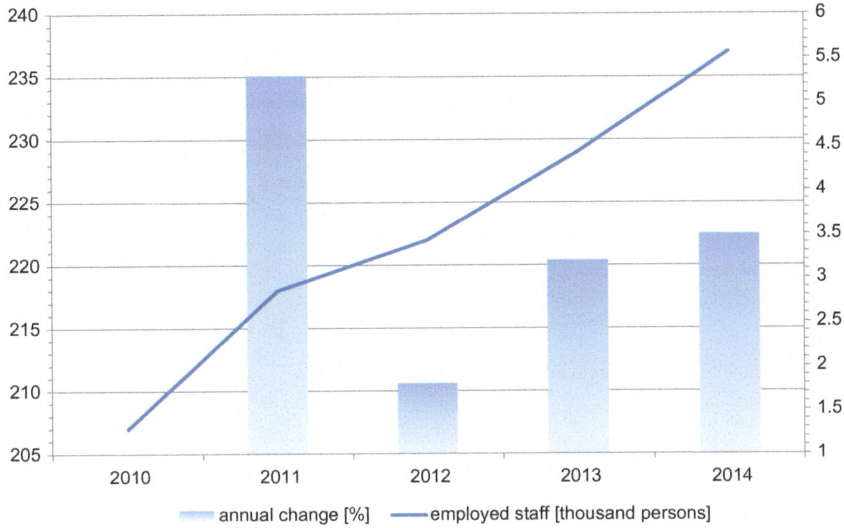

Fig. 6.10 German direct investment in the UK, 2010–2014, employment, employed staff left axis [thousand persons], annual change right axis [%] (*Source* Own elaboration on basis of Bundesbank [2017d])

The average annual growth rate for employment was 3.45% in the studied period, while average employment was 222,600. Considering jobs created by suppliers and other related enterprises, the overall figure was even greater. Adopting a leverage factor of 0.5–1.5 for additional employment effects, depending on industry, total employment provided by German investors in the UK both directly and indirectly was anywhere from 355,500 to 592,500 in 2014. The number of jobs offered in individual enterprises increased by 6.5% on average, from 162 to 173 per enterprise in 2010. This indicates that German investors not only increased their involvement in terms of the number of companies, but also invested substantially in staff needed for business activity. Average employment in the studied period was 167 workers per enterprise—fitting into the EU definition of small and medium-sized enterprises (Recommendation 2003/361/EG), meaning those that have fewer than 250 employees.

Fig. 6.11 German direct investment in the UK, 2010–2014, annual earnings, earnings left axis [EUR million], annual change right axis [%] (*Source* Own elaboration on basis of Bundesbank [2017e])

In terms of earnings, the development of investment was less steady than in other areas. As Fig. 6.11 shows, annual earnings increased by 22.3% from EUR 169.88 billion in 2010 to EUR 207.73 billion in 2014.[14] But the positive development trend was interrupted in 2012 when a drop of 3.6% was recorded. By contrast, in 2011 and 2014 the annual earnings growth rates hit the double digits. On average, earnings grew at an annual rate of 5.3%. Average earnings in the studied period were EUR 186.86 billion, while average annual earnings per enterprise increased by 13.7% from EUR 133.03 million in 2010 to EUR 151.29 million in 2014. Average annual earnings per enterprise in the analysed period stood at EUR 140.14 million.

[14]Bundesbank (2017e).

6.4 Potential Implications of Brexit for Bilateral Exchange and Conclusions

Brexit is bound to have serious consequences for bilateral economic relations between Germany and the UK. At the pre-Brexit stage, uncertainty is probably the most accurate term to describe the atmosphere in business today concerning future market and framework conditions on both sides of the English Channel. Since the Brexit referendum in 2016, many enterprises have redefined their investment plans and allocation decisions. Some have decided *ex ante* to adapt to any future changes. The importance of the City of London in the banking and finance sector is waning, and the city's position as the headquarters of international banks is visibly weakening. Many of them are thinking of moving to Frankfurt, which would strengthen Germany's position as a European and global banking hub. The development of the banking sector plays an important role in the public debate as well as in planning and calculations ahead of the UK's exit from the EU. While most investment decisions are long-term in nature, the exact conditions of Brexit remain unclear. Due to prolonged uncertainty, the situation after the Brexit vote has caused serious restrictions in planning and coordination for most enterprises involved in economic cooperation with the UK. The impact of Brexit on future investment projects is likely to be visible in the medium term. Especially, sectors and enterprises with significant exposure to business with Britain may be affected. Given the huge German investment in the UK, at EUR 128.6 billion, with EUR 21.7 million worth of capital invested per project on average, the potential risks of Brexit are evident. Meanwhile, the exposure of British investors in Germany is smaller, at EUR 29.7 billion, though the qualitative aspects of investment are also important. In recent years, a global trend towards a certain "flexibilisation" has been in evidence in investment, with producers moving their operations to other locations creating flexible production chains in response to changing market and framework conditions. With respect to the UK, many German investment projects, including those in core industries, can be qualified as flexible, allowing fast shifts of production segments back to the continent, like the assembly lines of the BMW Mini in Oxford Cowley, those of Bentley in Crewe and of Rolls Royce in Goodwood. But most German investment projects in Britain are strategic in nature and subject to adaption in the post-Brexit period. For the British economy, German investment is directly responsible for employing a quarter million people, with an average of 167 workers per company. A decline in investment after Brexit would have negative labour market implications at the aggregate level. Last but not least, the earnings of German-owned enterprises in the UK, at EUR 207.7 billion, are of substantial significance to the British economy. With average earnings at EUR 151 million per enterprise, German-owned businesses in Britain are in general large-scale enterprises, which is also important at the microeconomic level.

The prospects for bilateral trade relations will become clearer once the terms and conditions of Brexit are defined. The direction and extent of the impact will depend on future rules governing bilateral and multilateral trade flows, which are unknown at this point. The high level of Germany–UK trade, with EUR 86.2 billion worth of German exports and EUR 35.7 billion in imports, means that Brexit holds potential risks for both economies. Possible future tariffs and non-tariff barriers in the trade of goods may negatively affect traded volumes, depending on the price elasticities of demand and supply. Apart from supplier pricing policies, exchange rate developments will contribute to the final results of bilateral trade. The greatest risk of losses in the wake of Brexit-induced trade restrictions exists in three key sectors: machinery and transport equipment, whose exposure runs at EUR 47.2 billion; chemicals and related products, with a volume of EUR 12.8 billion; and manufactured products, with EUR 9.2 billion worth of exports. The UK is the No. 3 trade partner for Germany worldwide and the second largest in Europe, so compensating for shrinking trade volumes would be rather problematic. But because a major portion of the volumes is intra-industry trade any barriers will hamper productivity in the British economy by increasing costs and reducing availability. Serious losses can be expected.

German-British economic cooperation in bilateral trade and investment has largely been a success story in recent decades, both quantitatively and qualitatively. Even if the involvement of German partners is far more intense than on the British side in quantitative terms, any trade and investment restrictions after Brexit will negatively affect both economies. It is difficult to expect that any alternative free trade agreements, bilateral or multilateral, could compensate for potential losses caused by the UK's exit from the EU.

Appendix

See Figs. 6.12, 6.13, 6.14, 6.15, and 6.16.

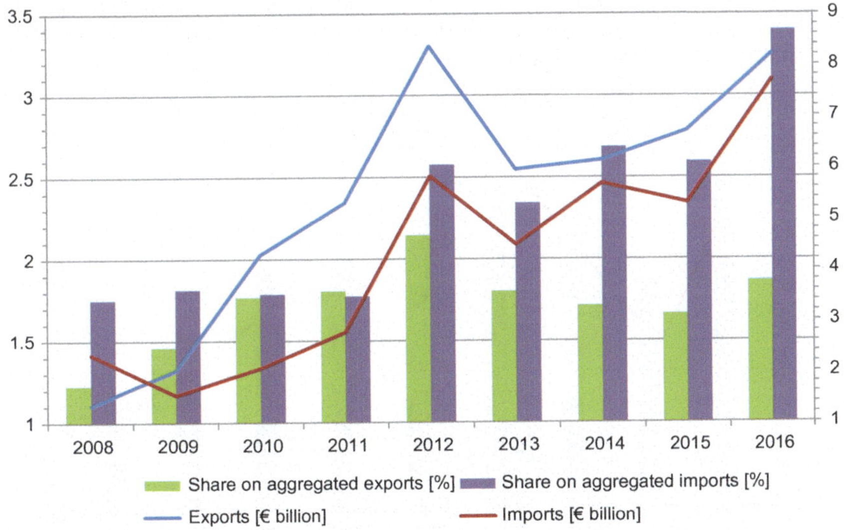

Fig. 6.12 Bilateral trade between Germany and the UK, 2008–2016, commodities and transactions not classified elsewhere in the SITC (SITC-9), left axis exports and imports [EUR billion], right axis shares of aggregated exports and imports [%] (*Source* Own elaboration on basis of Destatis [2017i])

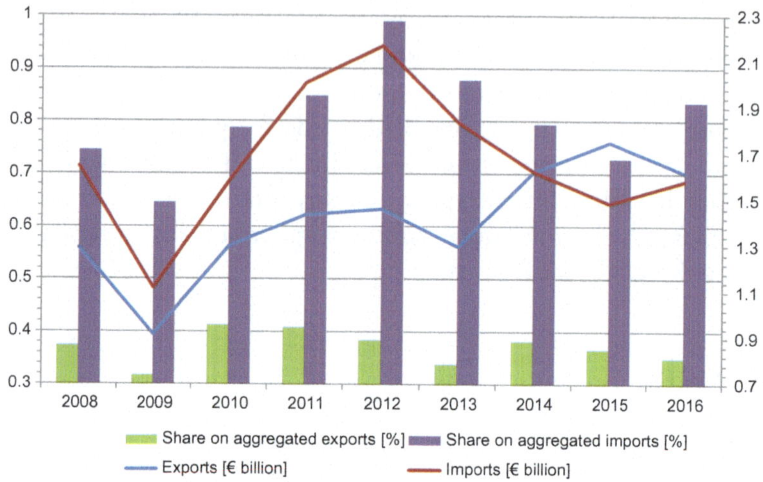

Fig. 6.13 Bilateral trade between Germany and the UK, 2008–2016, crude materials, inedible, except fuels (SITC-2), left axis exports and imports [EUR billion], right axis shares of aggregated exports and imports [%] (*Source* Own elaboration on basis of Destatis [2017j])

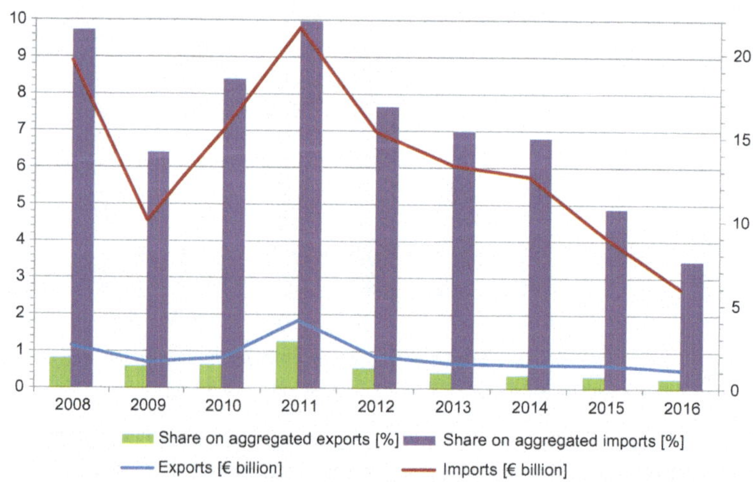

Fig. 6.14 Bilateral trade between Germany and the UK, 2008–2016, mineral fuels, lubricants and related materials (SITC-3), left axis exports and imports [EUR billion], right axis shares of aggregated exports and imports [%] (*Source* Own elaboration on basis of Destatis [2017k])

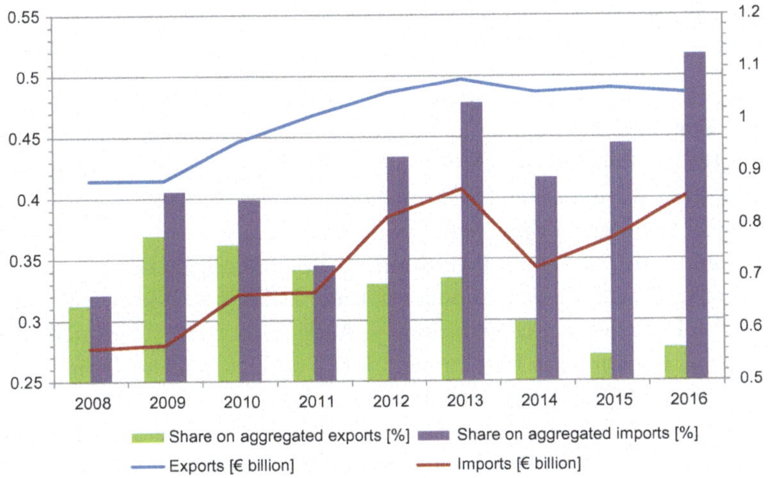

Fig. 6.15 Bilateral trade between Germany and the UK, 2008–2016, beverages and tobacco (SITC-1), left axis exports and imports [EUR billion], right axis shares of aggregated exports and imports [%] (*Source* Own elaboration on basis of Destatis [2017l])

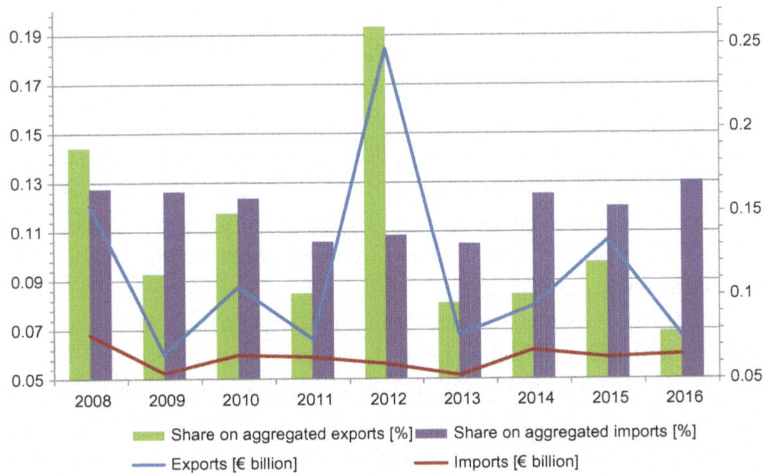

Fig. 6.16 Bilateral trade between Germany and the UK, 2008–2016, animal and vegetable oils, fats and waxes (SITC-4), left axis exports and imports [EUR billion], right axis shares of aggregated exports and imports [%] (*Source* Own elaboration on basis of Destatis [2017m])

References

Bundesbank. (2017a). Direktinvestitionen, Bestandsangaben, nach der alten nationalen Berechnungsmethodik bis Ende 2012, nach aktuellen Vorgaben der OECD Benchmark Definition, 4. Auflage ab 2010, Unmittelbare und mittelbare deutsche Direktinvestitionen im Ausland (konsolidiert) / Summe; Vereinigtes Königreich. bundesbank.de.

Bundesbank (2017b). Direktinvestitionen, Bestandsangaben, nach der alten nationalen Berechnungsmethodik bis Ende 2012, nach aktuellen Vorgaben der OECD Benchmark Definition, 4. Auflage ab 2010, Unmittelbare und mittelbare ausländische Direktinvestitionen in Deutschland (konsolidiert) / Summe; Vereinigtes Königreich. bundesbank.de.

Bundesbank. (2017c). Direktinvestitionen, Bestandsangaben, nach aktuellen Vorgaben der OECD Benchmark Definition, 4. Auflage ab 2010, Anzahl der Unternehmen; Vereinigtes Königreich. bundesbank.de.

Bundesbank. (2017d). Direktinvestitionen, Bestandsangaben, nach aktuellen Vorgaben der OECD Benchmark Definition, 4. Auflage ab 2010, Beschäftigte; Vereinigtes Königreich. bundesbank.de.

Bundesbank. (2017e). Direktinvestitionen, Bestandsangaben, nach aktuellen Vorgaben der OECD Benchmark Definition, 4. Auflage ab 2010, Jahresumsatz; Vereinigtes Königreich. bundesbank.de.

Destatis. (2017a). Genesis online, Außenhandelsstatistik, Aus- und Einfuhr (Außenhandel): Deutschland, Jahre, Länder, Vereinigtes Königreich, 1990–2016. www.genesis.destatis.de.

Destatis. (2017b). Außenhandel. Rangfolge der Handelspartner im Außenhandel der Bundesrepublik Deutschland 2016, Wiesbaden. destatis.de.

Destatis. (2017c). 12% der deutschen EU-Exporte gingen im Jahr 2016 in das Vereinigte Königreich. Pressemitteilung vom 5. Mai 2017, 147-17, Wiesbaden. www.destatis.de.

Destatis. (2017d). Genesis online, Außenhandelsstatistik, Aus- und Einfuhr (Außenhandel): Deutschland, Jahre, Länder, Vereinigtes Königreich, 2008–2016, SITC (1-Steller), Maschinen und Transport Equipment. www.genesis.destatis.de.

Destatis. (2017e). Genesis online, Außenhandelsstatistik, Aus- und Einfuhr (Außenhandel): Deutschland, Jahre, Länder, Vereinigtes Königreich, 2008–2016, SITC (1-Steller), Chemische Erzeugnisse. www.genesis.destatis.de.

Destatis. (2017f). Genesis online, Außenhandelsstatistik, Aus- und Einfuhr (Außenhandel): Deutschland, Jahre, Länder, Vereinigtes Königreich,

2008–2016, SITC (1-Steller), Bearbeit. Waren, vorwieg. n. Beschaffenheit geglied. www.genesis.destatis.de.
Destatis. (2017g). Genesis online, Außenhandelsstatistik, Aus- und Einfuhr (Außenhandel): Deutschland, Jahre, Länder, Vereinigtes Königreich, 2008–2016, SITC (1-Steller), Verschiedene Fertigwaren. www.genesis.destatis.de.
Destatis. (2017h). Genesis online, Außenhandelsstatistik, Aus- und Einfuhr (Außenhandel): Deutschland, Jahre, Länder, Vereinigtes Königreich, 2008–2016, SITC (1-Steller), Lebende Tiere und Nahrungsmittel. www.genesis.destatis.de.
Destatis. (2017i). Genesis online, Außenhandelsstatistik, Aus- und Einfuhr (Außenhandel): Deutschland, Jahre, Länder, Vereinigtes Königreich, 2008–2016, SITC (1-Steller), Waren und Warenverkehrsvorgänge. www.genesis.destatis.de.
Destatis. (2017j). Genesis online, Außenhandelsstatistik, Aus- und Einfuhr (Außenhandel): Deutschland, Jahre, Länder, Vereinigtes Königreich, 2008–2016, SITC (1-Steller), Rohstoffe (ausgen. Nahrungsm. u.mineral. Brennst). www.genesis.destatis.de.
Destatis. (2017k). Genesis online, Außenhandelsstatistik, Aus- und Einfuhr (Außenhandel): Deutschland, Jahre, Länder, Vereinigtes Königreich, 2008–2016, SITC (1-Steller), Mineral. Brennstoffe, Schmiermittel u.verw.Erzeug. www.genesis.destatis.de.
Destatis. (2017l). Genesis online, Außenhandelsstatistik, Aus- und Einfuhr (Außenhandel): Deutschland, Jahre, Länder, Vereinigtes Königreich, 2008–2016, SITC (1-Steller), Getränke und Tabak. www.genesis.destatis.de.
Destatis. (2017m). Genesis online, Außenhandelsstatistik, Aus- und Einfuhr (Außenhandel): Deutschland, Jahre, Länder, Vereinigtes Königreich, 2008–2016, SITC (1-Steller), Lebende Tiere und Nahrungsmittel. www.genesis.destatis.de.

7

Economic Consequences of Brexit for Poland

Mariusz-Jan Radło

7.1 Introduction

The process of the UK leaving the European Union is the consequence of a referendum in which Britons voted in favour of Brexit in June 2016. The decision triggered a debate on the implications of the exit for the UK, the EU and its individual member states as well as third countries. The debate has focused on a number of issues, including the macroeconomic impact of Brexit; its impact on specific markets and sectors (e.g. finance, energy, labour and innovation); international economic relations (e.g. bilateral and multilateral trade flows as well as FDI flows); EU policy (e.g. changes in voting powers in the EU Council, directions of European integration, the EU budget and agriculture policy); and other issues (e.g. migration, borders and the integrity of the UK).

All of these issues are likely to be shaped by Brexit. But no one knows exactly how the move will be implemented and what its final impact

M.-J. Radło (✉)
Collegium of World Economy,
SGH Warsaw School of Economics, Warsaw, Poland
e-mail: mjradlo@sgh.waw.pl

A. M. Kowalski (ed.), *Brexit and the Consequences for International Competitiveness*, https://doi.org/10.1007/978-3-030-03245-6_7

will be on post-Brexit relationships between the UK and the EU and its members. There are various scenarios concerning post-Brexit relationships between the EU and the UK. Barrett et al. (2015), just before the Brexit vote, indicated six possible ways in which these relations could be organised. Those were: the Norwegian scenario—the UK enjoying membership of the European Economic Area (EEA) and the European Free Trade Area (EFTA); the Swiss scenario—a bilateral agreement between the UK and the EU, with the UK being a member of EFTA; the Turkish scenario—the UK having a customs union with the EU; a preferential trade scenario—a bilateral free trade agreement between the UK and the EU; and no preferential trade scenario—no preferential trade agreement between the UK and the EU and no most favoured nation tariffs for Britain, unlike in the case of other countries belonging to the World Trade Organization.

In each of these scenarios, economic and institutional links between the UK and the EU are different. As a result, the economic effects of each scenario may be different. Within each individual scenario, it is possible to imagine different variants that can have different effects on individual economies. As pointed out by Busch and Matthes (2016), the Norwegian scenario would mean that Britain would not be part of EU decision-making processes, but would enjoy a customs union and full access to the single market (four freedoms). There would be no tariffs on EU–UK trade, but free trade agreements would be renegotiated, accompanied by an increased cost of rules of origin and customs clearance as well as limited regulatory autonomy. The UK would make financial contributions to the EU, but have little impact on EU regulations. The Swiss scenario is almost the same. However, the UK's participation in the single market would be more limited as the movement of goods, persons and services would only be partially free. Moreover, Britain would have no impact on EU regulations. The Turkish scenario assumes a lack of impact on EU decision-making, membership of the customs union and no tariffs on EU–UK trade, in addition to renegotiation of free trade agreements and limited access to the single market—with only partial free movement of goods, lack of free movement of persons and capital and rules on services in line with the General Agreement on Tariffs and Trade (GATS). Moreover, the Turkish scenario is characterised by no trade costs due to rules of origin (RoO), no cost of customs clearance, partial regulatory autonomy, no influence on EU regulations and no financial contributions to the EU. Finally, the "no preferential trade scenario" is similar to the Turkish one except for the lack of a customs union resulting in the existence of tariffs on EU–UK trade, no single market preferences, costs of customs clearance and full regulatory autonomy.

In their research, Busch and Matthes take into account the fact that a future agreement between the EU and the UK will have a significant impact on the costs and benefits of Brexit. Brexit will bring greater costs than benefits resulting from limitations in trade and reduced movement of capital and people between the EU and the UK. The potential cost of the UK leaving the EU can also have an unpredictable knock-on effect on both the UK and its EU partners, including Poland.

During the analysis, we will describe the impact of Brexit on migration in terms of international trade, foreign direct investment and decision-making processes in the EU, including changes in voting powers and the powers of selected coalitions within the EU. This chapter is structured as follows. In the first part, following the introduction, we review the macroeconomic consequences of Brexit. In the second part, we analyse the consequences of Brexit for migrations and their economic consequences. We then examine the implications of Brexit for trade and FDI, and finally, we investigate changes in EU decision-making and policies.

7.2 Macroeconomic Consequences

There are diverse estimates of how Brexit will affect the British economy. According to Begg and Mushövel (2016), some research reports (Minford) indicate that Brexit will have a positive impact on Britain's GDP. Other reports (Lions, Open Europe) suggest that the impact will be either positive or negative, and still others (HM Treasury, CEP, Oxford Economics, NIESR, PwC) find that Britain's departure from the EU will have an outright negative effect on the country's GDP. However, most researchers agree that Brexit will lead to a long-term loss of GDP for the UK in comparison with the status quo scenario—UK membership of the EU and the single market. Major sources of differences include assumptions about the ease of access to the EU single market (full versus restricted) in terms of trade and investment. Various researchers also use different estimation methods. However, as noted by Kierzenkowski et al. (2016), uncertainty about the outcome of the referendum and the consequences of Brexit has started to weaken growth in the UK.

Being a member of the European Union has contributed to the economic prosperity of the UK. Meanwhile, uncertainty about the outcome of the referendum has already started to dampen growth in the country. There are fears that Brexit could be a major negative shock to the UK economy, with economic fallout in the rest of the OECD, particularly other European countries. In some respects, Brexit would be akin to a tax on GDP, imposing a persistent and rising cost on the economy that would not be incurred if the UK remained in the EU. The shock would be transmitted through several channels that would change depending on the time horizon. In the near term, the UK economy would be hit by tighter financial conditions and weaker confidence and, after the country's formal exit from the European Union, by higher trade barriers and an early impact of restrictions on labour mobility. By 2020, GDP would be over 3% smaller than otherwise (with continued EU membership), equivalent to a cost per household of GBP 2200 (in today's prices). In the longer term, structural impacts would take hold through the channels of capital, immigration and lower technical progress. In particular, labour productivity would be held back by a drop in foreign direct investment and a smaller pool of skills. The extent of foregone GDP would increase over time. By 2030, in a central scenario, GDP would be over 5% lower than otherwise, with the cost of Brexit equivalent to GBP 3200 per household (in today's prices). The effects would be greater under a more pessimistic scenario and remain negative even under the optimistic scenario. Brexit would also hold back GDP in other European economies, particularly in the near term, resulting from heightened uncertainty about the future of Europe. By contrast, continued UK membership of the European Union and further reforms of the single market would enhance living standards on both sides of the Channel.

The macroeconomic estimates of Brexit's impact on individual economies, including Poland, are also diverse and depend on both the intensity of economic ties between specific countries and the Brexit scenario. For example, Schoof (2015) used an ifo Institute model for analysing trade policy based on a static model of general equilibrium, and found that, depending on the Brexit scenario, there are various negative consequences of Brexit for all EU member states. Poland is not particularly vulnerable to the effects of Brexit compared to other economies such as Ireland, Luxembourg, Malta, Cyprus or Belgium. These economies, because of their closer links with the UK, will suffer more from Brexit. Poland would see its real income shrink by just 0.07% in the case of a soft exit (Norwegian or Swiss scenario), and by anywhere from 0.16 to 0.24% in the case of the "deep cut" or "isolation" scenarios ("no preferential trade" scenario). Under all these scenarios, Poland would be hit far less than the abovementioned countries and slightly less than Germany.

7.3 Brexit and Migration Economics

Although migration from Poland to the UK appeared to be one of the reasons Britons said yes to Brexit (Kundera 2017), it has in fact brought more benefits to the UK than to Poland. As shown in Table 7.1, in 2016, there were 911,000 Polish-born nationals resident in the UK, up from 94,000 in 2004, when Poland joined the European Union. The almost tenfold increase was a consequence of accelerated labour migration from Poland to the UK, fostered by factors including tensions on the Polish labour market, low language barriers associated with growing English language proficiency, liberal UK policies towards immigrants, labour market difficulties in Poland, differences in wage levels between Poland and UK, and the depth and flexibility of the UK labour market.

As noted by Atoyan et al. (2016), migration from Poland and other countries that have joined the EU since 2004 has been dominated by educated and young people, which initially led to positive outcomes for

Table 7.1 Number of UK residents with Polish citizenship and born in Poland, 2004–2016

Year	Poland				UK residents with Polish citizenship and born in Poland (thousands)
	Total employment (thousands)	Unemployment (thousands)	Unemployment (%)		
			Total	Long term	
2004	127,202	29,996	18	10.3	94
2005	128,907	2773	16.7	10.3	162
2006	13,220	23,094	12.2	7.8	265
2007	137,711	17,466	8.5	4.9	411
2008	140,372	14,738	6.7	2.4	504
2009	137,823	18,927	8.5	2.5	529
2010	141,069	19,547	9.3	3	540
2011	142,326	19,827	9.7	3.6	654
2012	14,172	21,368	10.1	4.1	658
2013	142,443	21,579	9.8	4.4	688
2014	145,634	18,252	8.1	3.8	790
2015	148,298	15,633	6.9	2.9	831
2016	151,777	13,352	5.5	2.2	911

Source Own compilation based on estimates from Poland's Central Statistical Office and the Annual Population Survey by the UK Office for National Statistics

the migrants themselves, for the host economy and for the EU economy as a whole. However, large-scale migration also undermined the foundations of long-term economic growth in Poland, as it deepened demographic problems by reducing the working-age population and limiting the reproductive capacity of the Polish population. Similar conclusions were presented by Simionescu et al. (2017), who found that the loss of human capital in Poland due to migration to the UK had a negative impact on Poland's economic growth from 2004 to 2015.

Moreover, as shown by Homoncik et al. (2017), while migration from Poland to the UK initially reduced the tension on the labour market during a period of high unemployment in Poland, today it is causing increasingly visible shortages in the labour market when the unemployment rate approaches its natural level. This observation is confirmed by data in Table 7.1. More than 800,000 Poles migrated to the UK from 2004 to 2016. At the same time, the number of jobless in

Poland fell by 1.66 million: from almost 3 million in 2004 to just over 1.3 million in 2016. However, the number of those employed increased by 2.46 million, from 12.72 million to 15.78 million. Meanwhile, the unemployment rate in Poland dropped from 18 to 5.2%, and the long-term unemployment rate fell from 10.3 to 2.2%. This means that the drop in unemployment in Poland was mainly due to increased employment in the Polish economy and not emigration. In this perspective, if Brexit stopped the Polish labour force from leaving the country, it could have a positive impact on Poland's labour market.

From the point of view of the needs of the Polish labour market and the long-term development needs of the Polish economy, a key challenge is to reverse migration trends between Poland and the UK (as well as the rest of the EU). Thus, the key issue during Brexit negotiations for Poland should be not so much to guarantee access to the UK labour market for Polish citizens, but rather to create an effective mechanism to encourage them to return to Poland. As shown in Table 7.2, Poles in the UK are predominantly employed in those industries where the largest numbers of jobs have been created in Poland in recent years, and where there is the highest demand for labour.

Seen from this perspective, Brexit seems to be an opportunity rather than a challenge for Poland (see also Borońska-Hryniewiecka 2016). Both the rapid growth of the Polish economy and the prospects of its future development are conducive to exploiting the opportunity presented by Brexit to reverse migration trends. However, as pointed out by Atoyan et al. (2016), using this opportunity requires even better adjustment of policies to create an environment encouraging people to stay, promoting return migration and attracting skilled workers from other countries, including by strengthening institutions and economic policies. Poland also needs further improvements in workforce utilisation by increasing employment and productivity, accompanied by a mutual increase of wages to attract return migration. Moreover, better leveraging remittances accompanying migration could help promote investment rather than consumption. When shaping EU policies, Poland should support initiatives aimed at mitigating the negative effects of migration on growth and convergence.

Table 7.2 Poles employed in UK vs. vacancies and new jobs created in Poland by industry

Industry	Poles employed in UK		Vacancies and new jobs in Poland	
			Vacancies (2016)	New jobs created
	(Number)	(% total)	(% total)	(% total, 2009–2016)
Agriculture, forestry and fishing	11,092	1.7	n.a.	n.a.
Manufacturing	**156,863**	**23.5**	**27.0**	**20.9**
Electricity, gas, air conditioning supply	1030	0.2	n.a.	n.a.
Water supply, sewerage, waste	8033	1.2	n.a.	n.a.
Construction	**60,882**	**9.1**	**7.9**	**10.7**
Wholesale, retail, repair of vehicles	**98,089**	**14.7**	**18.2**	**26.0**
Transport and storage	**66,331**	**10**	**8.3**	**5.9**
Accommodation and food services	**64,258**	**9.6**	**4.3**	**4.4**
Information and communication	15,459	2.3	7.8	4.4
Financial and insurance activities	7305	1.1	2.9	1.8
Real estate activities	2868	0.4	n.a.	n.a.
Professional, scientific, technical activities	**22,696**	**3.4**	**7.9**	**6.8**
Administration and support services	**54,871**	**8.2**	**3.3**	**5.5**
Public administration and defence	6649	1	4.7	2.3
Education	16,779	2.5	1.8	5.5
Health and social work	44,260	6.6	4.1	3.5

(continued)

Table 7.2 (continued)

Industry	Poles employed in UK		Vacancies and new jobs in Poland	
			Vacancies (2016)	New jobs created
	(Number)	(% total)	(% total)	(% total, 2009–2016)
Arts, entertainment and recreation	7729	1.2	0.6	1.0
Other service activities	15,013	2.3	1.0	1.4
Households as employers	6249	0.9	n.a.	n.a.

Source Own compilation based on estimates from Poland's Central Statistical Office and the UK Office for National Statistics

7.4 Trade Implications

Various researchers dealing with the effects of Brexit say it will have a negative impact on international trade. Michalik (2017) argues that, due to the prominent role of non-EU countries in UK foreign trade, the negative influence of Brexit on UK trade will be limited. However, Brexit can produce negative consequences for Polish exporters. Similar findings have been presented by Pawlas (2017) and Bilan et al. (2017).

It should be noted that the share of the UK in Poland's exports of services is similar to that in goods exports: 7% vs. 6.61%—see Tables 7.3, 7.4, and 7.5. In the case of trade in services, the UK plays the most prominent role in Polish exports of financial services; telecommunications, computer and information services; and personal, cultural and recreational services. This means that potential barriers to trade in services may be detrimental to turnover in the financial and IT industries. In the case of trade in goods, Polish exports to the UK are particularly high in the case of motor vehicles, trailers and semi-trailers; basic pharmaceutical products and pharmaceutical preparations; rubber and plastic products; machinery and equipment; electronic and optical products; and food products. It should be noted that the EU's external tariffs are especially high for vehicle and automotive components and

Table 7.3 Trade in services: Poland vs. UK and World, 2015

Type of services	World (Euro, millions)		UK (Euro, millions)		UK (World = 100)	
	Exports	Imports	Exports	Imports	Exports	Imports
Total	406,573	297,524	28,403	25,996	7.0	8.7
Manufacturing services on physical inputs	31,125	3578	1295	55	4.2	1.5
Transport	108,267	63,636	5385	3561	5.0	5.6
Travel	94,295	71,492	1986	3219	2.1	4.5
Construction	13,867	9498	531	246	3.8	2.6
Insurance and pension services	3802	6625	198	734	5.2	11.1
Financial services	6173	8924	1520	2373	24.6	26.6
Charges for the use of intellectual property n.i.e.	3757	21,947	280	2488	7.5	11.3
Telecommunications, computer and information	39,641	24,878	6635	2856	16.7	11.5
Other business services	91,085	72,329	9060	8265	9.9	11.4
Personal, cultural and recreational services	4113	6527	695	1898	16.9	29.1

Source Own calculations based on OECD

Table 7.4 Trade in goods: Polish exports and imports to and from the UK and World, 2016

	World (USD, billions)		UK (USD, billions)	
	Exports	Imports	Exports	Imports
	196.46	188.52	12.98	4.81
D01: Crop and animal production, hunting and related service activities	3.82	4.38	0.17	0.02
D02: Forestry and logging	0.17	0.11	0.00	0.00
D03: Fishing and aquaculture	0.03	1.06	0.00	0.03
D05: Mining of coal and lignite	0.66	0.66	0.01	0.00
D06: Extraction of crude petroleum and natural gas	0.07	7.13	–	0.00
D07: Mining of metal ores	0.11	0.80	0.00	0.01
D08: Other mining and quarrying	0.17	0.51	0.01	0.01
D10T32: Manufacturing [C]	187.04	168.54	12.54	4.56
D10: Food products	18.98	10.98	1.87	0.35
D11: Beverages	0.81	0.77	0.07	0.11
D12: Tobacco products	2.15	0.75	0.13	0.03
D13: Textiles	2.61	3.99	0.07	0.12
D14: Wearing apparel	4.97	5.74	0.12	0.07
D15: Leather and related products	1.88	2.81	0.06	0.05
D16: Wood and products of wood and cork, except furniture	3.63	1.24	0.38	0.00
D17: Paper and paper products	4.95	4.95	0.24	0.09
D18: Printing and reproduction of recorded media	0.09	0.12	0.00	0.00
D19: Coke and refined petroleum products	3.69	3.26	0.11	0.06
D20: Chemicals and chemical products	12.92	19.74	0.73	0.74
D21: Basic pharmaceutical products and pharmaceutical preparations	2.91	5.69	0.11	0.35
D22: Rubber and plastics products	11.46	8.69	0.64	0.27
D23: Other non-metallic mineral products	4.03	2.37	0.37	0.05
D24: Basic metals	8.97	11.74	0.54	0.10

(continued)

Table 7.4 (continued)

	World (USD, billions)		UK (USD, billions)	
	Exports	Imports	Exports	Imports
D25: Fabricated metal products, except machinery and equipment	0.15	0.21	0.00	0.00
D26: Computer, electronic and optical products	16.10	20.73	1.61	0.36
D27: Electrical equipment	15.33	11.05	1.12	0.16
D28: Machinery and equipment n.e.c.	12.22	15.05	0.56	0.41
D29: Motor vehicles, trailers and semi-trailers	28.90	20.89	2.27	0.89
D30: Other transport equipment	7.53	4.90	0.13	0.07
D31T32: Furniture, other manufacturing	12.91	5.21	0.94	0.12

Source Own calculations based on OECD

machines. This means that a potential increase in tariffs may influence a major portion of Poland's exports.

Apart from developments affecting trade tariffs and important to trade and wider international cooperation, companies may have to deal with various changes affecting business. For example, Taylor Wessing (2016) identified several potential areas that can be influenced by Brexit, which may in turn affect international business cooperation, including dispute resolution, intellectual property protection, data protection and financial flows as well as corporate law and employment regulations. Similarly, KPMG (2017) found that Brexit might affect companies through changes in "passporting" rights for financial service companies, immigration and employment, regulated professions (e.g. lawyers), data protection legislation, income tax, indirect taxation and access to financial services (interest rate and exchange rate fluctuations).

Table 7.5 Structure of trade in goods: Polish exports and imports to and from the UK, 2016

	UK (World = 100)		UK (Total = 100)	
	Exports	Imports	Exports	Imports
Total	6.61	2.55	100	100
D01: Crop and animal production, hunting and related service activities	4.51	0.42	1.33	0.39
D02: Forestry and logging	1.64	0.30	0.02	0.01
D03: Fishing and aquaculture	2.22	3.11	0.01	0.68
D05: Mining of coal and lignite	1.30	0.28	0.07	0.04
D06: Extraction of crude petroleum and natural gas	–	0.00	–	0.00
D07: Mining of metal ores	3.17	1.76	0.03	0.29
D08: Other mining and quarrying	3.22	1.77	0.04	0.19
D10T32: Manufacturing [C]	6.70	2.71	96.58	94.76
D10: Food products	9.87	3.17	14.44	7.22
D11: Beverages	8.64	14.00	0.54	2.25
D12: Tobacco products	6.14	4.32	1.01	0.67
D13: Textiles	2.55	2.98	0.51	2.47
D14: Wearing apparel	2.43	1.17	0.93	1.40
D15: Leather and related products	3.21	1.96	0.47	1.14
D16: Wood and products of wood and cork, except furniture	10.55	0.31	2.95	0.08
D17: Paper and paper products	4.79	1.82	1.83	1.87
D18: Printing and reproduction of recorded media	3.53	3.35	0.02	0.08
D19: Coke and refined petroleum products	3.04	1.98	0.86	1.34
D20: Chemicals and chemical products	5.67	3.75	5.65	15.39
D21: Basic pharmaceutical products and pharmaceutical preparations	3.94	6.12	0.88	7.24
D22: Rubber and plastics products	5.56	3.13	4.91	5.64
D23: Other non-metallic mineral products	9.17	2.02	2.85	0.99
D24: Basic metals	6.01	0.89	4.15	2.17

(continued)

Table 7.5 (continued)

	UK (World = 100)		UK (Total = 100)	
	Exports	Imports	Exports	Imports
D25: Fabricated metal products, except machinery and equipment	1.41	0.16	0.02	0.01
D26: Computer, electronic and optical products	9.97	1.73	12.37	7.45
D27: Electrical equipment	7.29	1.47	8.61	3.37
D28: Machinery and equipment n.e.c.	4.60	2.71	4.33	8.46
D29: Motor vehicles, trailers and semi-trailers	7.84	4.24	17.45	18.40
D30: Other transport equipment	1.79	1.51	1.04	1.54
D31T32: Furniture, other manufacturing	7.26	2.27	7.22	2.46

Source Own calculations based on OECD

7.5 Impact on FDI

FDI flows are another area that may be affected by Brexit. However, as indicated by Bombińska (2017), the exact impact of Brexit on foreign direct investment flows will depend on the final model of cooperation between the European Union and the UK after Brexit. Adopting solutions based on close cooperation and institutional links will have little impact on FDI. As shown in Table 7.6, the UK is in sixth place among the countries of origin for the inward FDI stock in Poland, accounting for 5.1% of the total inward FDI stock. This means that small changes in regulations concerning FDI may be practically invisible in the UK's FDI in Poland. Meanwhile, greater regulatory changes resulting from Brexit may cause more visible changes in FDI flows. However, indirect UK-Poland links should also be taken into account. This is because the origins of FDI in Poland reflect the location of holding companies run by international corporations and not investment from specific countries. As shown by Schoof (2015), the UK is tightly linked to other locations friendly to international corporations, such as Luxembourg and the Netherlands. A less FDI-friendly form of Brexit could induce

Table 7.6 Structure of total inward FDI positions in Poland and total Polish outward FDI positions, 2016

Economy	Total inward FDI positions		Economy	Total outward FDI positions	
	Euro, billions	World = 100		Euro, billions	World = 100
Total World	176	100	Total World	27.67	100
Netherlands	33.9	19.3	Luxembourg	10.00	36.1
Germany	29.2	16.6	Cyprus	3.39	12.2
Luxembourg	23.5	13.3	Switzerland	2.06	7.4
France	17.8	10.1	Czech Republic	1.92	6.9
Spain	10.3	5.8	Netherlands	1.84	6.7
United Kingdom	**8.9**	**5.1**	Hungary	1.29	4.7
Italy	7.3	4.2	Canada	1.24	4.5
Austria	7.1	4.0	Germany	1.24	4.5
Belgium	6.2	3.5	Lithuania	1.00	3.6
Cyprus	5.9	3.3	United States	0.83	3.0
Switzerland	4.8	2.7	**United Kingdom**	**0.83**	**3.0**
United States	4.4	2.5	Russia	0.68	2.5
Sweden	3.5	2.0	Malta	0.68	2.5
Denmark	3.0	1.7	France	0.66	2.4
Finland	1.4	0.8	Romania	0.56	2.0
Ireland	1.3	0.7	Slovakia	0.39	1.4
Norway	1.2	0.7	Norway	0.34	1.2
Portugal	1.0	0.6	Chile	0.31	1.1
Japan	0.7	0.4	Turkey	0.27	1.0

Source Own compilation based on National Bank of Poland data

shifts in the origin of inward FDI. From the perspective of outward FDI, UK-based companies are not the most important market. Its share in total outward FDI from Poland is only 3%.

7.6 Impact on EU Decision-Making and Policies

From the perspective of the Polish economy, what matters is not only the influence of Brexit on GDP, trade, FDI and migration, but also the move's impact on EU decision-making and policies. Most studies point to a potential increase in Poland's role in the Council of the European Union.

However, it has also been pointed out that, as a result of Brexit, Poland will lose an ally supporting its approach to EU economic affairs and the directions of European integration. For example, Borońska-Hryniewiecka et al. (2017) observed that Poland and the UK shared similar views on EU economic policy issues such as support for deregulation in the EU common market, support for liberal EU trade policies and a common vision of integration. However, the UK did not share Poland's positive approach to cohesion policy. As a result, Brexit may strengthen hidden protectionist tendencies in the EU, contribute to over-regulated markets (e.g. services and labour), and lead to an unfavourable evolution of European integration from the point of view of Poland's European policy objectives.

Kóczy (2017) argued that Brexit would have a profound impact on the distribution of power in the Council of the European Union. Since the Lisbon Treaty the success of a voting initiative has exclusively depended on how many member states supported it and how many EU citizens lived in these countries in total. Using the Shapley–Shubik power index, Kóczy calculated the powers of member states with and without the UK and the latest population projections. He found that Brexit increased the power of the largest members while decreasing that of the smallest member states. The impact of Brexit on voting power in the Council of the EU was also assessed by Szczypińska (2017). She analysed power indices (Banzhaf power index) to examine changes in the distribution of power within the EU from the perspective of each EU member state separately as well as potential coalitions. She found that larger countries would benefit from the new power distribution while smaller nations would lose some of their power. According to Szczypińska, Poland would experience the highest relative increase in power (29%). The strength of various coalitions within the Council would also change. Euro-area member states would be able to adopt any proposal. However, Brexit will increase the ability of non-euro countries to build a blocking minority. The power of net beneficiaries after Brexit will increase in terms of forming a blocking minority, but the group of net contributors remains strong. Therefore, both coalitions can block any decision.

Meanwhile, Göllner (2017) analysed the impact of Brexit on the voting power of member countries in the European Council, a central hub of political decision-making in the EU. He also applied the Banzhaf power index and measured voting power before and after Brexit. He found that Brexit would result in an increase in the voting power of the Visegrad Group of Poland, the Czech Republic, Hungary and Slovakia compared to other European states. Moreover, among these states, Poland will see the biggest increase in its power index. Nevertheless, the real impact of the Visegrad Group will depend on its ability to coordinate voting behaviour.

Also worthy of note is a study by Pera (2017), who established that Brexit might deepen disintegration processes in the EU, and one by Hajdukiewicz (2017), who found that Brexit might influence factors driving Common Agricultural Policy (CAP) reform.

7.7 Conclusions

To sum up, Brexit may have diverse consequences for the Polish economy depending on the exact shape of economic relations between the EU and the UK after Brexit. Most macroeconomic analyses point to the negative effects of Brexit on GDP growth, but its actual impact on the Polish economy is expected to be negligible. In terms of migration, Brexit can have a positive impact on the Polish economy. Nevertheless, a key goal from the point of view of the interests of the Polish economy should be to use Brexit to reverse migration trends and to bring the labour force back to Poland. Thus, Brexit may positively influence Poland's labour market and its long-term economic growth. The impact of Brexit on trade and foreign investment, meanwhile, may be negative, though its actual influence may vary depending on the Brexit scenario. Interestingly, Brexit should improve the position of Poland and its regional Visegrad Group in the Council of the European Union and the European Council. It will also change the bargaining power of different coalitions of member states in terms of voting or blocking decisions within the EU.

References

Atoyan, M. R., Christiansen, L. E., Dizioli, A., Ebeke, M. C., Ilahi, M. N., Ilyina, M. A., et al. (2016). *Emigration and Its Economic Impact on Eastern Europe* (IMF Staff Discussion Note). Washington, DC: International Monetary Fund.

Barrett, A., Bergin, A., FitzGerald, J., Lambert, D., McCoy, D., Morgenroth, E., et al. (2015). *Scoping the Possible Economic Implications of Brexit on Ireland* (Research Series No. 48). Dublin: Economic and Social Research Institute.

Begg, I., & Mushövel, F. (2016). *The Economic Impact of Brexit: Jobs, Growth and the Public Finances*. London: London School of Economics and Political Science, European Institute.

Bilan, S., Kalyugina, S., & Mentel, G. (2017). Economic Implications of Brexit: Winners and Losers of European Divorce. *Czech Journal of Social Sciences Business and Economics, 6*(1), 6–12.

Bombińska, E. (2017). Brexit a bezpośrednie inwestycje zagraniczne w Wielkiej Brytanii. *Horyzonty Polityki, 8*(22), 171–192.

Borońska-Hryniewiecka, K. (2016). Wpływ Brexitu na sytuację imigrantów ekonomicznych w Wielkiej Brytanii: implikacje dla Polski i polskich obywateli. *Biuletyn PISM, 24*(1374), 1–2.

Borońska-Hryniewiecka, K., Kaca, E., Płóciennik, S., & Toporowski, P. (2017). *Relacje Unia Europejska—Wielka Brytania po Ewentualnym Brexicie Stanowiska Niemiec, Francji, Włoch, Hiszpanii i Polski*. Warszawa: Polski Instytut Spraw Międzynarodowych.

Busch, B., & Matthes, J. (2016). *Brexit—The Economic Impact: A Meta-Analysis* (IW-Reports from Institut der deutschen Wirtschaft Köln (IW)/ Cologne Institute for Economic Research, No. 10/2016). Köln.

Göllner, R. T. (2017). The Visegrád Group—A Rising Star Post-Brexit? Changing Distribution of Power in the European Council. *Open Political Science, 1*, 1–6.

Hajdukiewicz, A. (2017). Kierunki zmian Wspólnej Polityki Rolnej Unii Europejskiej w kontekście negocjacji handlowych na forum WTO i Brexitu. *Horyzonty Polityki, 8*(22), 109–130.

Homoncik, T., Pujer, K., & Wolańska, I. (2017). *Ekonomiczno-społeczne aspekty migracji: Wybrane problemy*. Wrocław: Wydawnictwo Exante.

Kierzenkowski, R., Pain, N., Rusticelli, E., & Zwart, S. (2016). *The Economic Consequences of Brexit: A Taxing Decision* (OECD Economic Policy Paper Series). Paris: OECD.

Kóczy, L. A. (2017). *How Brexit Affects European Union Power Distribution*. Accessed May 18, 2016. Available at SSRN. https://ssrn.com/abstract=2781666 or http://dx.doi.org/10.2139/ssrn.2781666.

KPMG. (2017). *Brexit—wyzwaniem dla międzynarodowych i polskich przedsiębiorstw*. Warszawa: KPMG.

Kundera, J. (2017). Brexit. *Economics of the 21st Century, 2*(14), 56–72.

Michalik, A. (2017). Brexit a wymiana handlowa Wielkiej Brytanii. Wnioski dla Polski. *Horyzonty Polityki, 8*(22), 91–107.

Pawlas, I. (2017). *Powiązania gospodarcze między Polską i Wielką Brytanią w kontekście Brexitu* (Studia Ekonomiczne—Zeszyty Naukowe Uniwersytetu Ekonomicznego w Katowicach, No. 319/2017, pp. 169–183).

Pera, B. (2017). Brexit a wymiana handlowa Unii Europejskiej i Wielkiej Brytanii. *Horyzonty Polityki, 8*(22), 69–89.

Schoof, U. (2015). *Costs and Benefits of a United Kingdom Exit from the European Union*. Gütersloh: Bertelsmann Stiftung.

Simionescu, M., Bilan, Y., & Mentel, G. (2017). Economic Effects of Migration from Poland to the UK. *Amfiteatru Economic, 19*(46), 757–770.

Szczypińska, A. (2017). *Who Gains More Power in the EU After Brexit?* (MF Working Paper Series). Warsaw: Ministry of Finance, Republic of Poland.

Taylor Wessing. (2016). *Brexit: Practical Consequences for Entrepreneurs from Poland and the UK*. Warsaw: Taylor Wessing.

8

Economic Implications of Brexit for the International Competitiveness of Russia

Krzysztof Falkowski

8.1 Introduction

The ongoing process of the UK leaving the European Union provides a good opportunity to consider to what extent, if at all, the UK's exit from the European Union (widely known as "Brexit") will affect its relations with Russia and, in particular, the competitiveness of the Russian economy. While the conditions of this "divorce" remain unclear, one thing is certain: the UK will still be a member of many international organisations after Brexit, including the World Trade Organisation, which will have consequences for its foreign trade policy. By exiting the EU, the UK will regain its autonomy in shaping mutual economic relations with third countries, including Russia.

The main aim of this chapter is to outline the possible economic consequences of Brexit for the international competitive position of the

---

K. Falkowski (✉)
Collegium of World Economy,
SGH Warsaw School of Economics, Warsaw, Poland
e-mail: kfalkow@sgh.waw.pl

A. M. Kowalski (ed.), *Brexit and the Consequences for International Competitiveness*, https://doi.org/10.1007/978-3-030-03245-6_8

Russian economy. The first part of the study provides a concise analysis of Russia's competitiveness in international trade considering four basic categories of goods, as classified by the Organisation for Economic Cooperation and Development (OECD) based on their technological advancement. For this purpose, two indicators are applied: Balassa's Revealed Comparative Advantage (RCA) index and the Lafay Index (LFI) of international trade specialisation. The analysis is conducted for the 2000–2016 period to ensure that the identified comparative advantages can be considered on a long-term basis. Subsequently, in the following three sections of this chapter, bilateral economic relations between Russia and the UK in the fields of trade, investment and labour migration are discussed in a concise way depending on the availability of data. In each section, the potential effects of Brexit on the competitiveness of the Russian economy are examined.

This chapter puts forth a thesis that, due to the extent of mutual economic cooperation and the competitive profile of the Russian economy, Brexit will have little influence on Russia's competitiveness in the world economy. Moreover, as shown by the limited effects of Western sanctions imposed on Russia in 2014, it seems that political relations between Russia and the UK will be far more important to Russia's future competitive position than the UK's upcoming exit from the EU.

8.2 Russia's Competitive Profile in Contemporary International Trade

Economists use a variety of different methods to assess the competitiveness of economies in international trade, which is understood as the ability to achieve greater benefits (than other countries) from both their own and foreign factors of production under the conditions of an open economy (Weresa 2014). This includes the ability to develop, produce and sell products and services that are more attractive in terms of price or quality than those exported by other countries. As a result, a country plays an increasing role in the trade of such goods internationally (Carbaugh 2017). Methods used to assess the international competitiveness of economies have been reviewed by researchers including Startiene and Remeikiene (2014).

In this chapter, the following two indicators are used to assess the competitiveness of the Russian economy in international trade: Balassa's Relative Comparative Advantage (RCA) index (1965, 1989) and the LFI of international trade specialisation (1992).

The values of the first indicator were determined based on the logarithmic form of the original formula by Balassa (1965, 1989), according to the following formula:

$$\mathrm{RCA}_{ij}^{K} = \ln\left(\frac{\frac{x_{ij}^{K}}{X_{j}^{K}}}{\frac{x_{i}^{j}}{X^{j}}}\right)$$

where:

RCA_{ij}^{K}—the RCA index of country K in goods category i as compared to country j or group of countries j

x_{ij}^{K}—exports of goods category i from country K to country j or group of countries j

X_{j}^{K}—total exports from country K to country j or group of countries j

x_{i}^{j}—exports of goods category i from country j or group of countries j

X^{j}—total exports from country j or group of countries j

i—goods category

K—analysed country

j—other countries (rest of the world).

The use of the logarithmic form of the above formula allows for the symmetry of the positive and negative RCA_{ij}^{K} indices in the range around zero, which facilitates their interpretation (Falkowski 2017a). One can say that a country possesses a RCA in trade in goods category i only if the share of this category in the country's total exports is higher than the share of goods category i in total global exports, so when $\mathrm{RCA}_{ij}^{K} > 0$.

The value of the second indicator, the LFI of international trade specialisation (1992)—which is widely used to assess the nature of a country's foreign trade balance and indirectly also its international competitiveness—was determined in accordance with the following formula:

$$\mathrm{LFI}_{ij}^{K} = 100 \cdot \left(\frac{x_{ij}^{K} - m_{ij}^{K}}{x_{ij}^{K} + m_{ij}^{K}} - \frac{X_{j}^{K} - M_{j}^{K}}{X_{j}^{K} + M_{j}^{K}} \right) \cdot \frac{x_{ij}^{K} + m_{ij}^{K}}{X_{j}^{K} + M_{j}^{K}}$$

where:

LFI_{ij}^{K}—the LFI of country K in goods category i as compared to country j or group of countries j

x_{ij}^{K}—exports of goods category i from country K to country j or group of countries j

X_{j}^{K}—total exports from country K to country j or group of countries j

m_{ij}^{K}—imports of goods category i from country j or group of countries j

M_{j}^{K}—imports of country K from country j or group of countries j

i—goods category

K—analysed country

j—other countries (rest of the world).

In this case, one can say that a country's international trade shows comparative advantages when the value of the indicator for goods category i is positive ($\mathrm{LFI}_{ij}^{K} > 0$), which means that the country has a trade surplus for goods category i.

Based on the values of the RCA and LFI indices for the 2000–2016 period, Russia's competitiveness in international trade was evaluated within the four basic categories of goods according to the OECD classification based on their technological advancement. Such an approach makes it possible to assess the country's international competitiveness with respect to high-tech, medium-high-tech, medium-low-tech and low-tech goods (OECD 2011; Hatzichronoglou 1997).

When analysing the RCA and LFI indices (Fig. 8.1) for Russia, it can be observed that the country's competitiveness in international trade is low. The only RCAs it possesses are with respect to medium-low-tech goods, which include raw materials and their processed derivatives, in the trade of which Russia has consistently been highly competitive internationally (Fig. 8.2). In the case of the other three goods categories according to the OECD classification, i.e. high-tech, medium-high-tech and low-tech goods, Russia did not have any RCA in the analysed period, as evidenced by the negative values of both the RCA and LFI indices. Moreover, Russia is highly uncompetitive in the

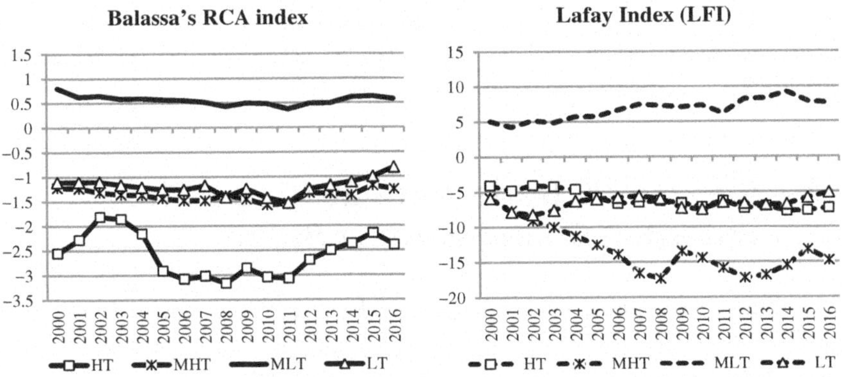

Fig. 8.1 Russia's RCA in the trade of high-tech, medium-high-tech, medium-low-tech and low-tech goods, 2000–2016 (HT—high-tech goods, MHT—medium-high-tech goods, MLT—medium-low-tech goods, LT—low-tech goods; *Source* Own elaboration based on data from the United Nations Commodity Trade Statistics Database)

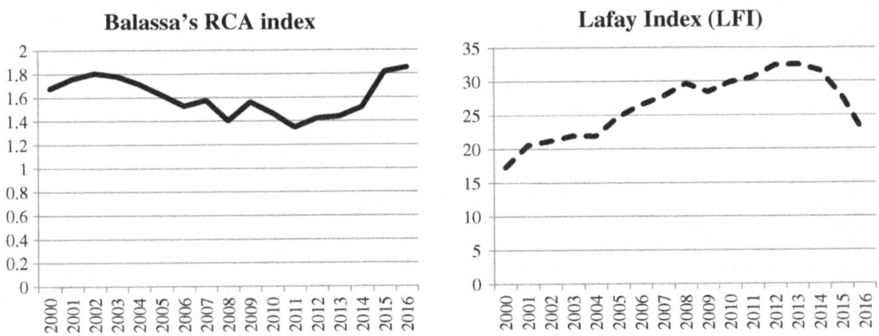

Fig. 8.2 Russia's RCA in the trade of mineral fuels, lubricants and related materials, 2000–2016 (*Source* Own elaboration based on data from the United Nations Commodity Trade Statistics Database)

trade of high-tech goods. From the point of view of the specific features of contemporary international trade and the growing role of high-tech goods (Wu et al. 2017), this fact should be viewed as particularly worrying not only with regard to Russia's future role in the world economy but also the country's further development. It is necessary to mention the so-called resource curse affecting Russia (Falkowski 2013, 2017b), meaning the country's exclusive international specialisation in the trade of mineral fuels and their derivatives. This makes the country vulnerable to price fluctuations on international markets.

8.3 Russia–UK Trade from 2010 to 2016; Potential Changes After Brexit

The volume of trade between Russia and the UK was subject to significant fluctuations during the analysed period of 2010–2016 (Fig. 8.3). While bilateral trade initially grew from USD 14.3 billion in 2010 to USD 20.7 billion in 2012 (an increase of 44.8%), from 2013 onwards it started falling, at first slightly, then dramatically. In 2013, total trade turnover was USD 20.5 billion; by the end of 2016 it had shrunk to USD 10.4 billion, declining by nearly 50%. A key reason was sanctions that the EU, including the UK, imposed on Russia in 2014 in connection with the annexation of Crimea and the beginning of a conflict in eastern Ukraine where Ukrainian separatists were formally supported by

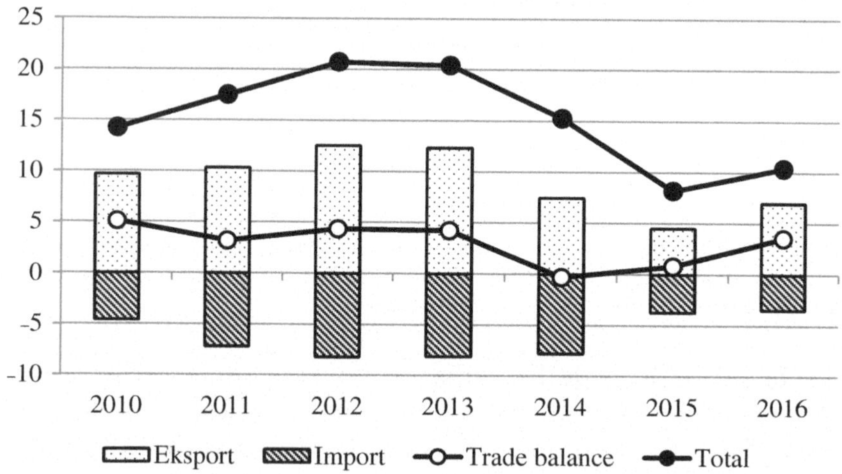

Fig. 8.3 Russia's trade with the UK, 2010–2016 (USD billion) (*Source* Own elaboration based on data from the United Nations Commodity Trade Statistics Database)

Russia. Another, more important, reason was Russian counter-sanctions imposed on EU member states.

As a result, Russia's trade balance with the UK deteriorated dramatically in 2014 and 2015 (Fig. 8.3). While in 2008 the positive balance of trade was USD 7.3 billion, followed by USD 5.1 billion in 2010, in 2014 the value of Russian imports from the UK exceeded that of Russian exports to that country for the first time in the twenty-first century. Consequently, Russia recorded a negative trade balance with the UK at around USD 300 million. In 2015, the situation improved slightly for Russia, with a positive balance of around USD 800 million.

In 2016, as shown in Fig. 8.3, Russian exports to the UK and total trade turnover as well as the trade balance were all up on the corresponding values for 2015. However, the same cannot be said of the value of Russian imports from the UK, which fell by about USD 3 billion. Russia's imports from the UK began declining in 2013.

The structure of Russian exports to the UK from 2010 to 2016—within the four basic goods categories classified by the OECD based on technological advancement—shows that medium-low-tech goods figured most prominently among Russia's exports, especially the *Coke, refined petroleum products and nuclear fuel* subcategory (Table 8.1). However, the value of such exports dropped sharply from USD 4.7 billion in 2010 to USD 1.4 billion in 2016. Meanwhile, the value of Russian medium-low-tech exports in the *Basic metals and fabricated metal products* subcategory increased from USD 0.3 billion in 2010 to USD 1.1 billion in 2016.

Russia exports a marginal volume of high-tech goods to the UK. Under the OECD classification, these include the following subcategories: *Aircraft and spacecraft*; *Medical, precision and optical instruments*; *Office, accounting and computing machinery*; *Pharmaceuticals*, and *Radio, TV and communications equipment.* Such exports peaked at around USD 120 million in 2012 and 2013, but then fell again, largely because Russian high-tech goods are uncompetitive on international markets.

Russian imports from the UK in the 2010–2016 period (Table 8.2) were dominated by medium-high-tech goods, especially *Motor vehicles, trailers and semi-trailers* and *Machinery and equipment, n.e.c*; and high-tech goods, especially *Pharmaceuticals* and *Medical, precision and optical*

Table 8.1 Russia's exports to the UK by type of goods, according to the OECD classification, 2010–2016 (USD million)

Category	Subcategory	2010	2011	2012	2013	2014	2015	2016
HT	Aircraft and spacecraft	9.9	11.3	85.3	82.2	45.0	14.1	8.4
	Medical, precision and optical instruments	13.5	22.8	22.1	25.6	29.1	24.4	21.0
	Office, accounting and computing machinery	2.7	1.4	1.3	3.8	4.0	7.0	5.2
	Pharmaceuticals	0.6	0.3	1.2	0.3	3.1	1.3	1.5
	Radio, TV and communications equipment	9.3	9.7	9.1	10.0	8.4	9.2	6.9
	HT Total	**35.9**	**45.5**	**119.0**	**121.9**	**89.6**	**56.1**	**42.9**
MHT	Chemicals excluding pharmaceuticals	310.7	188.2	209.5	120.7	150.9	142.6	102.4
	Electrical machinery and apparatus, n.e.c.	13.4	6.4	5.9	16.2	14.2	17.3	8.9
	Machinery and equipment, n.e.c.	20.6	11.5	15.5	23.3	19.6	28.7	21.4
	Motor vehicles, trailers and semi-trailers	0.6	0.4	1.7	2.8	1.8	3.6	4.6
	Railroad equipment and transport equipment, n.e.c.	0.1	0.2	0.2	0.4	0.4	0.6	1.1
	MHT Total	**345.4**	**206.6**	**232.8**	**163.4**	**186.8**	**192.8**	**138.2**
MLT	Basic metals and fabricated metal products	297.2	593.1	4158.7	840.3	894.3	404.4	1099.1
	Building and repairing of ships and boats	2.8	8.9	–	0.1	–	0.4	56.9
	Coke, refined petroleum products and nuclear fuel	4749.9	5568.5	3789.2	6060.8	4007.0	2561.8	1426.6
	Other non-metallic mineral products	8.5	6.4	5.7	5.9	7.2	8.9	16.7
	Rubber and plastics products	1.2	1.8	1.9	4.5	5.4	11.2	10.8
	MLT Total	**5059.5**	**6178.7**	**7955.5**	**6911.6**	**4914.0**	**2986.7**	**2610.1**

(continued)

Table 8.1 (continued)

Category	Subcategory	2010	2011	2012	2013	2014	2015	2016
LT	Food products, beverages and tobacco	60.2	71.3	68.7	76.6	70.0	53.4	36.6
	Manufacturing, n.e.c.	4.9	4.9	8.8	5.4	6.9	6.1	5.6
	Textiles, textile products, leather and footwear	4.4	2.4	7.3	4.5	2.6	17.0	5.6
	Wood, pulp, paper, paper products, printing and publishing	142.6	152.2	134.3	126.2	159.8	150.3	143.2
	LT Total	**212.2**	**230.7**	**219.0**	**212.7**	**239.3**	**226.9**	**190.9**
	Other Total	**4042.7**	**3688.4**	**4035.6**	**4945.2**	**2074.1**	**1061.3**	**4014.8**
	Export Total	**9695.7**	**10,349.9**	**12,561.8**	**12,354.7**	**7503.8**	**4523.7**	**6996.9**

Source Own elaboration based on data from the United Nations Commodity Trade Statistics Database

Table 8.2 Russia's imports from the UK by type of goods, according to the OECD classification, 2010–2016 (USD million)

Category	Subcategory	2010	2011	2012	2013	2014	2015	2016
HT	Aircraft and spacecraft	0.7	30.8	62.1	6.2	1.4	17.5	–
	Medical, precision and optical instruments	230.7	248,9	337.3	305.9	295.0	174.3	148.2
	Office, accounting and computing machinery	36.1	26.4	27.5	19.8	14.4	8.5	10.3
	Pharmaceuticals	533.6	626.8	606.2	816.6	490.5	433.2	418.7
	Radio, TV and communciations equipment	46.6	64.0	71.2	56.0	228.5	33.7	46.9
	HT Total	**847.8**	**996.9**	**1104.3**	**1204.5**	**1029.7**	**667.2**	**624.1**
MHT	Chemicals excluding pharmaceuticals	551.6	946.1	1048.1	890.3	697.1	530.6	552.8
	Electrical machinery and apparatus, n.e.c.	252.9	332.6	420.5	411.4	1161.8	336.0	160.5
	Machinery and equipment, n.e.c.	578.6	1181.0	1392.5	1228.1	956.1	463.7	517.8
	Motor vehicles, trailers and semi-trailers	1167.5	2280.0	2697.7	2761.3	2107.6	808.6	726.2
	Railroad equipment and transport equipment, n.e.c.	0.5	2.2	2.3	3.5	4.3	2.6	1.4
	MHT Total	**2551.1**	**4741.9**	**5561.1**	**5294.6**	**4926.9**	**2141.4**	**1958.7**
MLT	Basic metals and fabricated metal products	227.1	277.5	273.6	236.5	281.6	150.1	129.8
	Building and repairing of ships and boats	9.2	36.5	40.6	59.0	29.3	26.0	3.7
	Coke, refined petroleum products and nuclear fuel	76.4	67.0	82.6	64.0	64.1	45.7	38.1
	Other non-metallic mineral products	39.5	47.8	55.5	72.4	66.9	34.8	34.7
	Rubber and plastics products	140.9	202.5	210.0	199.6	184.4	116.2	122.9
	MLT Total	**493.1**	**634.4**	**662.4**	**631.5**	**626.2**	**372.8**	**329.3**

(continued)

Table 8.2 (continued)

Category	Subcategory	2010	2011	2012	2013	2014	2015	2016
LT	Food products, beverages and tobacco	399.0	517.3	582.1	601.3	543.6	314.1	306.0
	Manufacturing, n.e.c.	35.4	43.4	38.7	41.7	50.0	46.3	36.9
	Textiles, textile products, leather and footwear	38.2	47.0	62.8	75.7	64.8	35.5	27.8
	Wood, pulp, paper, paper products, printing and publishing	175.0	132.7	113.6	213.7	509.1	102.7	76.1
	LT Total	**647.6**	**740.3**	**797.2**	**932.4**	**1167.4**	**498.6**	**446.7**
	Other Total	**27.8**	**49.4**	**66.6**	**43.5**	**27.8**	**20.0**	**73.4**
	Import Total	**4567.4**	**7162.8**	**8191.6**	**8106.4**	**7778.0**	**3700.1**	**3432.2**

Source Own elaboration based on data from the United Nations Commodity Trade Statistics Database

instruments. At the end of the analysed period, the value of Russian imports of such goods decreased, leading to a decline in the country's overall imports from the UK.

If we look at the balance of trade between Russia and the UK from 2010 to 2016 (Fig. 8.4), it is clear that Russia had a positive trade balance only in medium-low-tech goods, a category in which it consistently possessed RCAs (as demonstrated in the first part of this study). Likewise, Russia's trade balance was positive in crude oil and natural gas. This positive trade balance declined markedly in 2014–2015, due to a smaller volume of trade in these goods rather than a significant decrease in world oil prices.

Russia repeatedly recorded a negative trade balance for the other three OECD categories, i.e. high-tech, medium-high-tech and low-tech goods. The trade balance in medium-high-tech goods was especially unimpressive, especially from 2011 to 2014. Meanwhile, an overall decline in the value of bilateral trade led to a "flattening" of the trade balances in all three categories in 2015–2016.

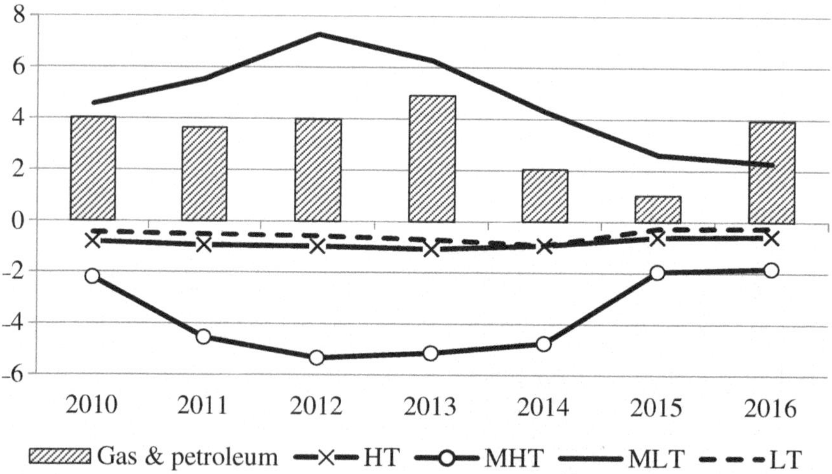

Fig. 8.4 Balance of Russia–UK trade in individual goods categories, according to an OECD classification based on technological advancement, as well as in gas and oil, 2010–2016 (USD billion) (*Source* Own elaboration based on data from the United Nations Commodity Trade Statistics Database)

In analysing the structure of Russia's trade with the UK, a different perspective can be taken with a classification of goods according to the UN Standard International Trade Classification Rev. 4 or SITC Rev. 4 (Tables 8.3 and 8.4).

Taking into account all Russian exports to the UK in 2010–2016, as shown in Table 8.3, *Mineral fuels, lubricants and related materials* (Section 3) dominated until 2015, accounting for 85.5% of Russia's total exports to the UK in the peak year of 2013. In 2016, *Commodities and transactions not classified elsewhere in the SITC* (Section 9) represented 49% of total Russian exports to Britain.

As reflected by the data, the structure of Russian exports to the UK showed no significant diversification during the analysed period. Except for Sections 3 and 9 as well as *Manufactured goods classified chiefly by material* (Section 6), all the remaining categories played an insignificant role in exports. The share of goods from Section 6 increased more than threefold over the analysed period, from 3.8% in 2010 to 12.2% in 2016.

Meanwhile, Russian imports from the UK (Table 8.4) were dominated by *Machinery and transport equipment* (Section 7), whose share ranged from 43.1% of total imports in 2016 to 58.7% in 2014. *Chemicals and related products, n.e.s.* (Section 5) also played a significant role, accounting for 15.8% of total imports in 2014 and 29.4% in 2016.

This analysis of the structure of Russia's trade with the UK shows that Russian exports were dominated by low-value-added and labour-intensive goods rather than capital-intensive and low-cost products. The key role was played by *Mineral fuels, lubricants and related materials* (Category 3), goods in which Russia is unquestionably competitive internationally, as evidenced by its strong RCAs.

On the other hand, Russian imports from the UK were dominated by highly processed, high-value-added and capital-intensive goods rather than labour-intensive, medium-high-tech products—mainly *Machinery and transport equipment* (Section 7).

To sum up, the structure of Russian–UK trade by type of goods should be described as unfavourable for the Russian economy, chiefly because of unfavourable terms of trade and a strong dependence on the

Table 8.3 Russian exports to the UK by type of goods, according to the SITC Rev. 4 classification, 2010–2016 (%)

Code	Section	2010	2011	2012	2013	2014	2015	2016
0	Food and live animals	0.2	0.3	0.2	0.2	0.4	0.7	0.6
1	Beverages and tobacco	0.1	0.2	0.2	0.2	0.5	0.5	0.3
2	Crude materials, inedible, except fuels	2.3	1.1	0.7	1.1	3.8	3.6	1.6
3	**Mineral fuels, lubricants and related materials**	**74.5**	**85.4**	**57.7**	**85.5**	**78.2**	**77.7**	**32.7**
4	Animal and vegetable oils, fats and waxes	0.3	0.3	0.2	0.3	0.2	0.2	0.0
5	Chemicals and related products, n.e.s.	2.1	2.0	1.9	1.1	2.0	3.3	1.6
6	Manufactured goods classified chiefly by material	3.8	6.6	12.4	6.7	6.4	10.2	12.2
7	Machinery and transport equipment	0.6	0.6	1.0	1.1	1.3	1.8	1.6
8	Miscellaneous manufactured articles	0.3	0.3	0.3	0.3	0.5	1.1	0.5
9	Commodities and transactions not classified elsewhere in the SITC	15.8	3.4	25.4	3.4	6.7	0.9	49.0
Total (%)		100.0	100.0	100.0	100.0	100.0	100.0	100.0

Source Own elaboration based on data from the United Nations Commodity Trade Statistics Database

Table 8.4 Russia's imports from the UK by commodity according to the SITC Rev. 4 classification, 2010–2016 (%)

Code	Section	2010	2011	2012	2013	2014	2015	2016
0	Food and live animals	3.6	2.4	1.9	2.2	2.2	2.4	2.8
1	Beverages and tobacco	5.3	5.2	5.5	5.4	5.0	6.3	6.4
2	Crude materials, inedible, except fuels	0.8	0.7	0.6	0.5	0.4	0.6	0.6
3	Mineral fuels, lubricants and related materials	1.3	1.0	1.0	0.8	0.8	1.3	0.7
4	Animal and vegetable oils, fats and waxes	0.0	0.0	0.0	0.0	0.0	0.0	0.0
5	Chemicals and related products, n.e.s.	24.5	19.7	16.9	19.9	15.8	26.8	29.4
6	Manufactured goods classified chiefly by material	10.2	7.2	6.3	6.1	6.7	8.0	8.0
7	**Machinery and transport equipment**	**47.5**	**55.8**	**58.6**	**56.4**	**58.7**	**46.3**	**43.1**
8	Miscellaneous manufactured articles	6.6	8.0	8.9	8.6	10.4	8.3	7.4
9	Commodities and transactions not classified elsewhere in the SITC	0.1	0.0	0.3	0.1	0.0	0.0	1.6
Total (%)		100.0	100.0	100.0	100.0	100.0	100.0	100.0

Source Own elaboration based on data from the United Nations Commodity Trade Statistics Database

prices of energy raw materials on international markets. Moreover, the structure of Russian–UK trade closely matches the competitive profile of the Russian economy. Russia exports to the UK goods in which it possesses RCAs (medium-low-tech goods), while importing those in which it has no RCAs (especially medium-high-tech goods).

As it remains unclear what rules will govern the UK's commercial relations with other countries in the future, it is difficult to clearly identify the possible consequences of Brexit for Russia's competitiveness in trade with Britain. However, considering that Russia and the UK are not key trading partners for each other, Brexit should not affect the international competitiveness of the Russian economy. This assertion seems all the more justified since Russia's biggest RCAs in international trade are in energy raw materials, weapons and military aircraft of which the UK is not a significant importer. This would mean that any potential changes in trade conditions for these goods after Brexit are unlikely to affect competitiveness. On the other hand, Russia imports ready-made, highly processed medium-high-tech industrial goods that do not add to the country's competitiveness.

8.4 Investment Cooperation Between Russia and the UK in 2010–2015; Potential Changes After Brexit

Another important area of economic relations between Russia and the UK is investment cooperation. The mutual importance of Russia and the UK as investment partners has been relatively high for years. The UK, alongside the Netherlands, Cyprus and Germany, has traditionally been a major investor in Russia, while Russia is a major investor in the UK where Russian capital is mostly invested in real estate and shares of companies listed on the London Stock Exchange.

When analysing mutual direct investment (Fig. 8.5) and equity investment (Fig. 8.6) in 2010–2015 (with 2015 being the latest year for

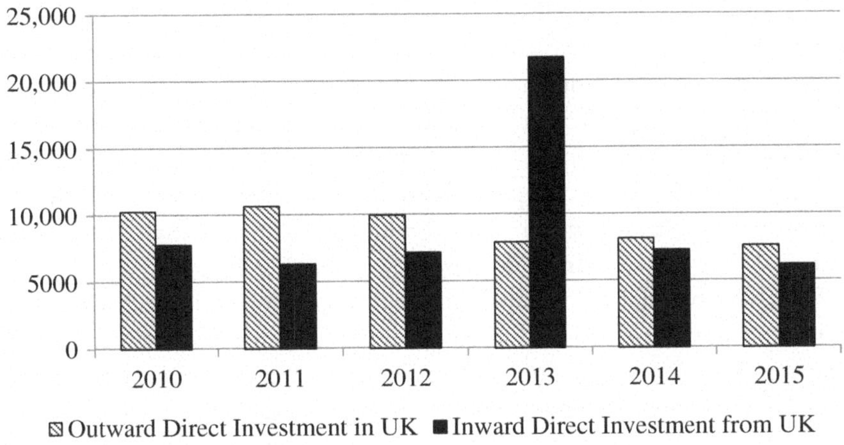

Fig. 8.5 Russia–UK direct investment, 2010–2015 (USD million) (*Source* Own elaboration based on the International Monetary Fund data)

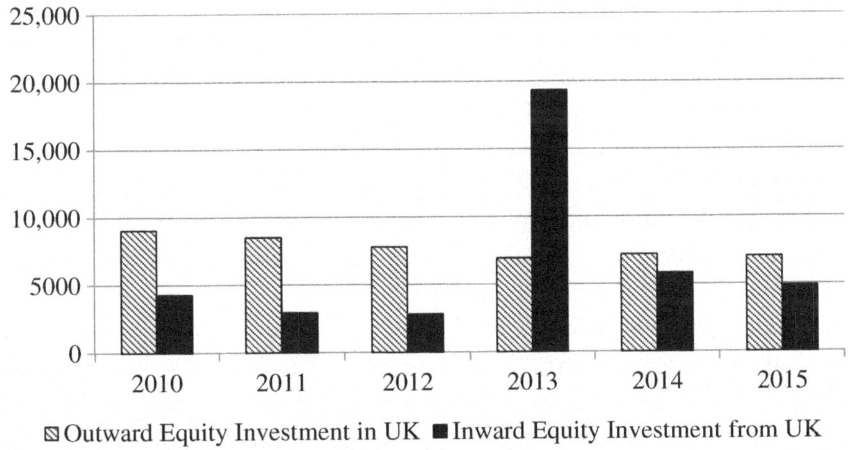

Fig. 8.6 Russia–UK equity investment, 2010–2015 (USD million) (*Source* Own elaboration based on the International Monetary Fund data)

which IMF data is available), it should be noted that Russian investment in the UK was consistently higher than British investment in Russia. The only exception was 2013 when British direct and equity investment in Russia was nearly three times as high as Russian investment in Britain. The spike in UK investment in Russia that year was due to an IPO of oil company Rosneft as well as investment by the British-based Royal Dutch Shell group in projects including LNG (Russia Direct 2017; Santander 2016).

In 2014 and 2015, UK investment flows into Russia declined, especially compared to 2013, due to geopolitical tensions between Russia and the West over Ukraine and an economic crisis (decreased attractiveness of shares of Russian energy companies as a result of falling oil prices on international markets). For instance, in the aftermath of EU sanctions imposed on Russia, Royal Dutch Shell had to stop working with Gazpromneft in a shale oil development project. Consequently, in 2015 the UK was ranked a distant 11th among the largest foreign investors in Russia, with 2.9% of total FDI.

As regards capital flows (both direct and portfolio investments) between Russia and the UK, no major changes should be expected post-Brexit. The countries' bilateral political relations will be far more important (e.g. the question of keeping Western sanctions against Russia or Russian counter-sanctions in place) than the UK's exit from the European Union unless there is a major decline in the attractiveness of the UK economy and Russian investors lose interest in investing in Britain. But such a scenario is practically impossible due to the strength and importance of the UK economy.

Significantly, UK investment in the extraction of Russian energy raw materials, particularly natural gas, and sales of technology for extracting deeply buried gas deposits, decreased markedly as a result of Western sanctions imposed on Russia (Falkowski 2015). This undoubtedly had a negative effect on Russia's competitiveness given its comparative advantages in the extraction and export of energy raw materials (Falkowski 2013, 2017b) .

8.5 Migration Between Russia and the UK from 2010 to 2016; Potential Changes After Brexit

Another important dimension of relations between countries is the flow of labour. Recently, its significance has grown tremendously in both Europe and elsewhere.

In the case of relations between Russia and the UK, there is a clear imbalance in this area. As shown in Fig. 8.7, the UK is a far more attractive destination to settle and find a job for Russians than Russia is for Britons. Considering the total migrant stock, the number of Russians who officially emigrated to the UK increased from 25,439 in 2005 to 42,491 in 2015 (up by 67%). To compare, the number of British citizens who immigrated to Russia rose from 1267 in 2005 to 1741 in 2015 (an increase of 37.5%).

Likewise, no significant changes should be expected in terms of population migration after Brexit. The current visa regime is highly likely to continue in the future, while labour market access will be regulated

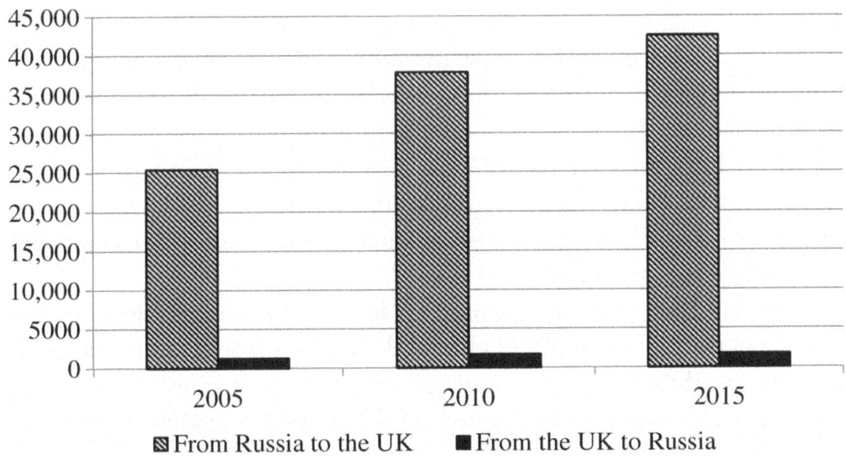

Fig. 8.7 Russia–UK total migrant stock in 2005, 2010 and 2015 (number of people) (*Source* Own elaboration based on United Nations data)

by bilateral agreements. Therefore, the countries' bilateral political relations will be of key importance. Nevertheless, the growing migration of young and well-educated Russians to the UK may have a negative impact on Russia's competitiveness.

8.6 Conclusions

The international competitiveness of the Russian economy is low and does not reflect the country's potential. This is a direct consequence of the policy of abandoning transformation and modernisation in the economic and social system inherited from the former USSR. Russia's competitive profile in international trade is predominately based on mineral resources, especially energy, which makes the economy dependent on volatile price developments in international commodities markets. As shown by the analysis in this chapter, Russia possesses RCA only in the trade of medium-low-tech goods, which include raw materials and their processed derivatives. Russia does not have any advantages in low-tech, medium-high-tech and high-tech goods.

In view of the above, the impending exit of the UK from the EU should have little effect on Russia's international competitiveness. Although Russian–UK trade is dominated by Russian exports of mineral fuels, lubricants and related materials, Britain is not a significant importer of such goods. On the other hand, Russia imports from the UK ready-made, highly processed medium-high-tech industrial goods that meet domestic demand but do not add to the competitiveness of the Russian economy. It does not seem reasonable to expect that the existing commodity structure of mutual trade could change anytime soon, not even in the medium term.

Brexit will have even less impact on investment cooperation between Russia and the UK, which will chiefly depend on macroeconomic developments in these countries and government policy towards foreign investment. In this context, a potential significant change in the economic situation in the UK could affect the country's investment attractiveness and thus Russian investment, especially portfolio investment. But such a scenario has no basis in fact today. In Russia, in turn, much will depend not only on the macroeconomic situation but also on the Kremlin's policy towards foreign investors in sectors such as raw materials and mineral resources, which are key to the country's international competitiveness. Nor should any serious changes be expected in population migration between Russia and the UK as a result of Brexit.

In conclusion, due to the scale of mutual economic cooperation and the competitive profile of the Russian economy, Brexit should have little impact on Russia's international competitiveness. Much more important for Russia's future competitiveness will be macroeconomic factors, but, above all, efforts to modernise its economy and reorient it towards one based on knowledge and technology rather than raw materials.

References

Balassa, B. (1965). Trade Liberalization and 'Revealed' Comparative Advantage. *Manchester School of Economic and Social Studies, 33*(2), 99–123.

Balassa, B. (1989). 'Revealed' Comparative Advantage Revisited. In B. Balassa (Ed.), *Comparative Advantage, Trade Policy and Economic Development* (pp. 63–79). New York: New York University Press.

Carbaugh, R. (2017). *International Economics*. Boston: Cengage.

Falkowski, K. (2013). *Międzynarodowa konkurencyjność gospodarek Białorusi, Rosji i Ukrainy*. Warszawa: Szkoła Główna Handlowa.

Falkowski, K. (2015). Bezpieczeństwo ekonomiczne Rosji w warunkach sankcji gospodarczych UE w kontekście posiadanych międzynarodowych przewag konkurencyjnych. *Studia Europejskie, 4*(76), 153–173.

Falkowski, K. (2017a). Innovativeness of Belarusian Economy Through the Prism of Its Competitive Position in International Trade. *CBU International Conference Proceedings 2017, 5*, 120–127.

Falkowski, K. (2017b). Long-Term Comparative Advantages of the Eurasian Economic Union Member States in International Trade. *International Journal of Management and Economics, 53*(4), 27–49.

Hatzichronoglou, T. (1997). *Revision of the High Technology Sector and Product Classification* (STI Working Papers 1997/2, OECD/GD 97[216]). Paris: Organisation for Economic Cooperation and Development.

Lafay, G. (1992). The Measurement of Revealed Comparative Advantages. In M. G. Dagenais & P. A. Muet (Eds.), *International Trade Modeling*. London: Chapman & Hill.

OECD. (2011). *ISIC Rev. 3 Technology Intensity Definition: Classification of Manufacturing Industries into Categories Based on R&D Intensities*. http://www.oecd.org/sti/ind/48350231.pdf. Accessed October 13, 2017.

Russia Direct. (2017, October 22). *Russia-EU Economic Relations: Assessing Two Years of Sanctions*. http://www.russia-direct.org/analysis/russia-eu-economic-relations-assessing-two-years-sanctions. Accessed October 22, 2017.

Santander. (2016). https://en.portal.santandertrade.com/establish-overseas/russia/foreign-investment. Accessed October 22, 2017.

Startiene, G., & Remeikiene, R. (2014). Evaluation of Revealed Comparative Advantage of Lithuanian Industry in Global Markets. *Procedia—Social and Behavioral Sciences, 110*, 428–438.

Weresa, M. A. (Ed.). (2014). *Innovation, Human Capital and Trade Competitiveness: How Are They Connected and Why Do They Matter?* Switzerland: Springer International Publishing.

Wu, J., Ma, Z., & Zhuo, S. (2017). Enhancing National Innovative Capacity: The Impact of High-Tech International Trade and Inward Foreign Direct Investment. *International Business Review, 26*(3), 502–514.

9

Trade Openness and FDI in the UK After Brexit

Tomasz M. Napiórkowski

9.1 Introduction

The June 2016 referendum and the decision of UK citizens to leave the European Union sent shock waves through Europe and posed a number of economic risks to Britain. The uncertainty coming from this event and the wide range of possibilities for the exit procedure itself will see their reflection as economic shocks both, locally and globally.

The aim of this chapter is to examine how exiting the EU will impact the attractiveness of the UK to foreign direct investment (FDI).[1]

[1]Defined by UNCTAD, the source of data, as "an investment made to acquire lasting interest in enterprises operating outside of the economy of the investor. Further, in cases of FDI, the investor's purpose is to gain an effective voice in the management of the enterprise… [with]… a threshold of 10% of equity ownership to qualify an investor as a foreign direct investor" (UNCTAD 2017a).

T. M. Napiórkowski (✉)
Collegium of World Economy,
SGH Warsaw School of Economics, Warsaw, Poland
e-mail: tnapio@sgh.waw.pl

A. M. Kowalski (ed.), *Brexit and the Consequences for International Competitiveness*, https://doi.org/10.1007/978-3-030-03245-6_9

The hypothesis of the study is that the UK will be a less attractive destination for foreign investors after Brexit.

The novel aspect of this study is that it frees itself of the assumptions on the final rules of the UK leaving the EU, which is needed to establish a precise cost of the departure. The topic is approached from a theoretical perspective. This results in a set of conclusions about expected changes. However, the magnitude of these changes is heavily scenario-dependent and goes beyond the scope of this study, which focuses on the theoretical aspects of the examined problem.

This chapter is structured as follows. First, a study of literature on the impact of FDI on the host economy and on the determinants of FDI will be conducted. This will show the types of benefits that the UK may lose due to a retreat of FDI and will make it possible to establish the role of trade openness as a key determinant of FDI. Second, the status quo of inward FDI in the UK will be presented, with the aim of showing the scale of inward FDI activity in the UK. Third, data analysis (with the use of the Pearson linear correlation coefficient and Granger causality) will be conducted as a means of showing the relationship between inward FDI and trade openness in the UK.

9.2 The Benefits of Hosting and the Determinants of Inward FDI: Literature Study

The aim of this review is twofold: to show the benefits of hosting FDI (summarised in Fig. 9.1)—which, at least to some degree, will be foregone in the event of a retreat of inward FDI from the UK; and to show the overview of the determinants of inward FDI and the key role of trade openness within that set.

The direct benefit of FDI in the short term is an increase in the level of domestic investment, which translates into a higher level of capital and therefore faster economic growth. Although research reports on the relationship between inward FDI and domestic investment contain the argument that FDI is crowding out domestic investment, the overall results appear to be mixed. Based on a study by Pilbeam and

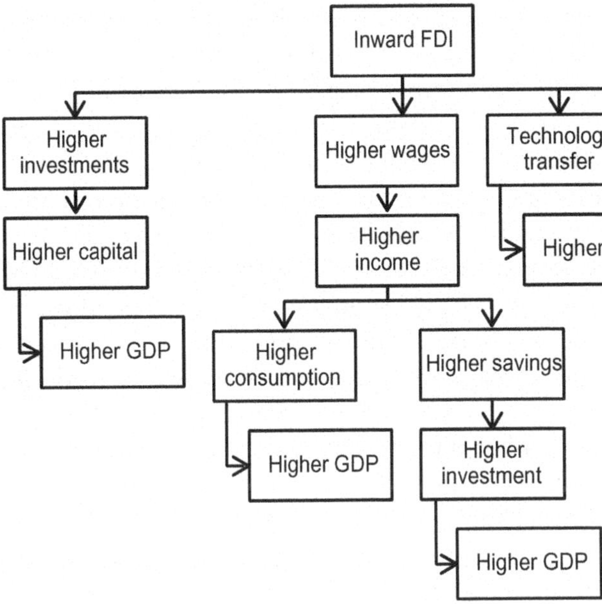

Fig. 9.1 Benefits of inward FDI to the host economy (*Source* Author's own)

Oboleviciute (2012), a conclusion can be reached that the more developed the host is, the higher is the magnitude of the crowding out of domestic investment. Interestingly, when studying a sub-group of the subjects considered by Pilbeam and Oboleviciute (2012), Szkorupová (2015) reported quite opposite findings. The divergence of results on the studied relationship is highlighted by an exchange between Morrissey and Udomkerdmonkol (2012, 2016), and Farla et al. (2016). Authors such as Ahmed et al. (2015) show that, while for the entire economy the crowding out of domestic investment can be of no significance, this may not be true when studying sector by sector. Another benefit highlighted by the researchers is that foreign firms tend to pay higher wages than their domestic counterparts (Lipsey 2002; Tomohara and Takii 2011; Javorcik 2015). These higher wages, through the marginal propensity to save, are transformed into higher saving, which become additional investments. Given the presence of sufficient absorptive capacity (Nunnenkamp 2002a; Velde 2006; Azam and Ahmed 2015), FDI also brings with it technology (Liu et al. 2016; Svedin and

Stage 2016) and an increase in human capital via know-how transfer (Branstetter 2006; Tülüce and Doğan 2014; Temiz and Gökmen 2014).

Shifting the topic to the determinants of FDI, the literature is filled with an abundance of possible determinants of FDI and their various permutations (see discussion by Blonigen and Piger 2011). Therefore, it is impossible to declare some determinants not important or rate their general economic importance (Przybylska 2001; Stawicka 2013) as these can be case-study dependent (Leitão 2010). Walsh and Yu (2010) show that the significance of FDI determinants will also depend on the classification of FDI, i.e. primary, secondary and tertiary. Further, the determinants of foreign direct investment can be classified according to the type of investment, e.g. horizontal versus vertical FDI (Navaretti and Vanables 2006). Despite these variations, some authors maintain that, despite the process of globalisation, traditional determinants remain important (Nunnenkamp 2002b). When examining reports on the determinants of FDI, a core set of inward FDI explanatory variables can be extracted. These are: (i) market size/economic potential/state (usually proxied by GDP and its per capita permutation and their growth—for example, see: Bevan and Estrin 2000; Nunnenkamp 2002a; Walsh and Yu 2010; Kalemli-Ozcan and Nikolsko-Rzhevskyy 2010; Barros et al. 2013; Napiórkowski 2015)[2,3]; (ii) labour cost (represented by, e.g. average weekly earnings, the difference between, or the ratio of, the wages in the host and in the home economy—for example, see: Narula and Wakelin 1997; OECD 2000; Kerr and Peter 2001; Carstensen and Toubal 2003; Napiórkowski 2014); and (iii) trade and its costs as a representation of trade openness (expressed, e.g. with various permutations on the export and import variables of the host,

[2]Adding to the topic, two authors, Carstensen and Toubal (2003), state that the market potential is related to markets present in neighbouring economies, in addition to the host's own domestic market. As a result, the researchers take into account the distance (i.e. proxy for the host's internal transportation costs) as well as the transportation cost between the recipient and the investing economy.

[3]The work of Xun and Awokuse (2005) is an interesting one as it uses comparative gross domestic product determinants, such as the squared difference between the two parties involved and the sum of the host's and home's values of this economic variable adding also an interaction term between the skill and gross domestic product differences.

tariffs, tariff revenues, distance between economies, trade agreements or custom unions—for example, see: Beer and Cory 1996; Buch et al. 2001; Nunnenkamp 2002b; Carstensen and Toubal 2003; Blonigen and Piger 2011; Napiórkowski 2013).

This overview of the literature has shown that every decrease in inward FDI in the UK will impact its economy via a number of channels. Some of these effects, e.g. employment and wages, will be seen in the short term, while others, e.g. changes in capital accumulation, will be seen in the medium term, and still others, e.g. changes in the growth rate for technology, will be seen in the long term. A fall in these elements will translate into one of the chief determinants of FDI, namely GDP, a fall in which will result in an additional loss of attractiveness of inward FDI, especially market-seeking FDI. A fall in capital will chiefly impact decisions related to resource-seeking FDI, and a fall in the growth rates for technology and human capital will translate into a fall of inward FDI-seeking strategic assets.

The read-through of the literature on the determinants of FDI has also shown that, whereas the "tail" determinants are selected on a case-by-case basis, there are three key explanatory factors that are always used in one form or another. One of these is the openness of the home economy to trade, which usually carries a hypothesis that the bigger is the openness of the economy to trade, the higher are the inflows of FDI to that economy. This mirrors the base of the research hypothesis of this study. However, a second, opposite, scenario must also be considered, namely that a decrease in the ability to deliver goods and services to a given market with exports will force local production by a foreign firm, leading to an increased value of inward FDI.

9.3 FDI in the UK

The first aim of this section is to show the long-term inward FDI trends in the UK. Second, by examining inward FDI activity as a percentage of the UK's GDP, the importance of inward FDI to the UK economy will be presented. Third, using inward FDI expressed as a percentage of the world's total, the relative attractiveness of the UK as a destination for FDI

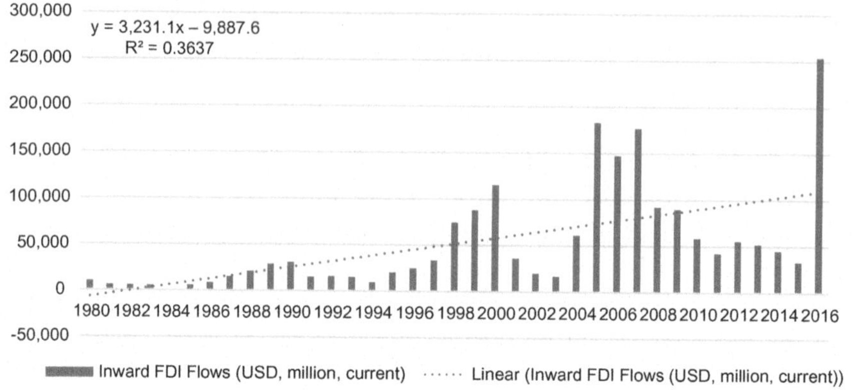

Fig. 9.2 Inward FDI flows in the UK in millions of current USD (*Source* Author's own based on data from UNCTAD [2017b])

will be analysed. Fourth, the sources of inward UK FDI will be analysed, which will show the dependence of inward FDI in the UK on the EU.[4]

Inward FDI flows[5] (Fig. 9.2) in the UK present an uneven trend that is related to global economic shocks, e.g. the bubble in the early 2000s and the international financial crisis, which started in late 2007 in the USA and at the end of the decade in Europe. Therefore, it is possible to state a hypothesis that a negative economic shock in the form of the UK leaving the EU will see its reflection in inward FDI flows to the UK. A surprise comes in the form of a significant outlier in the value of inward FDI flows in 2016 (USD 253,825.77 million).[6] The inward

[4]Given that the paper focuses on the UK as a recipient of FDI, only inward FDI activity will be analysed.

[5]"For associates and subsidiaries, FDI flows consist of the net sales of shares and loans (including non-cash acquisitions made against equipment, manufacturing rights, etc.) to the parent company plus the parent firm's share of the affiliate's reinvested earnings plus total net intra-company loans (short- and long-term) provided by the parent company. For branches, FDI flows consist of the increase in reinvested earnings plus the net increase in funds received from the foreign direct investor" (UNCTAD 2015a).

[6]This surge in inflows of FDI into the UK is, at the time of this study, treated as an outlier, as changes in the key macroeconomic determinants of FDI (e.g. GDP, labour costs) do not seem to justify such a significant increase. This treatment is further supported by the fact that the start of the Brexit procedure significantly increased the risk on undertaking FDI in the UK, which should

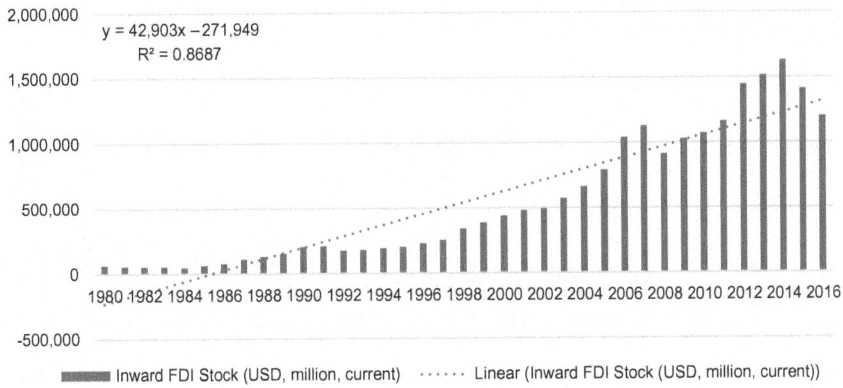

Fig. 9.3 Inward FDI stock in the UK in millions of current USD (*Source* Author's own based on data from UNCTAD [2017b])

FDI stock[7] (Fig. 9.3) has a clear positive trend, which appears to be breaking after reaching its peak in 2014 (USD 1,628,518.6 million).

Similar conclusions can be reached when examining the inward FDI flows and stock as a percentage of the UK's GDP (Fig. 9.4). The data on flows is significantly distorted by the 2016 value, which corresponds to 9.79%, while the average for the 1980–2016 period is 2.43%, close to the global total of 2.29%. The FDI stock's value as a share of the UK's GDP has been falling since 2013 (55.62%) and stood at 46.15% in 2016, above the world total of 35.07%. Therefore, it can be concluded that FDI as a building block is more significant in the UK that in the world on average (Fig. 9.5).

---

result in a fall in inward FDI activity. However, one must recognise that the shock of the decision to Brexit may not yet be fully represented in the available data as decisions to conduct FDI take time to make; hence, the shock of the referendum results may not have yet been calculated in the said decisions.

[7]"For associate and subsidiary enterprises, it is the value of the share of their capital and reserves (including retained profits) attributable to the parent enterprise (this is equal to total assets minus total liabilities), plus the net indebtedness of the associate or subsidiary to the parent firm. For branches, it is the value of fixed assets and the value of current assets and investments, excluding amounts due from the parent, less liabilities to third parties" (UNCTAD 2015b).

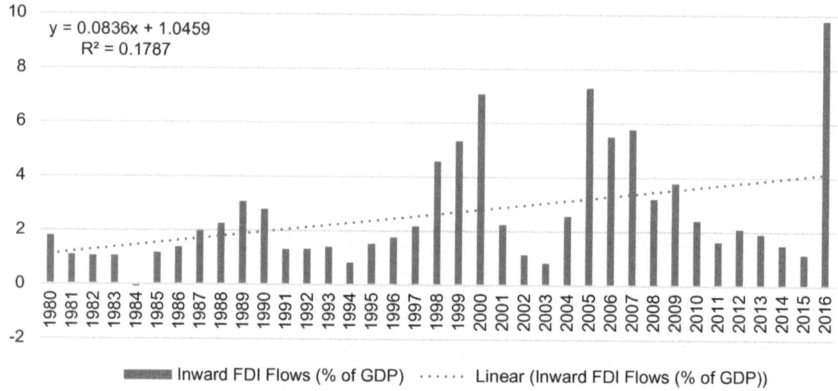

Fig. 9.4 Inward FDI flows in the UK as a percentage of UK GDP (*Source* Author's own based on data from UNCTAD [2017b])

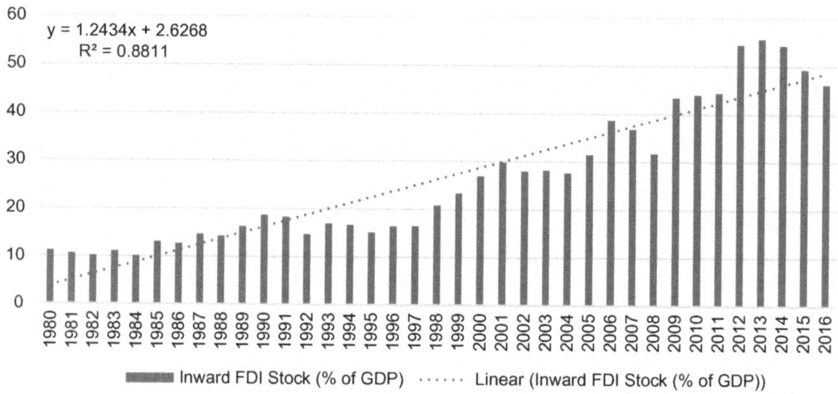

Fig. 9.5 Inward FDI stock in the UK as a percentage of UK GDP (*Source* Author's own based on data from UNCTAD [2017b])

The relative attractiveness of the UK as a destination of FDI flows has been almost steadily decreasing since 2005 when 19.08% of all world flows were directed at the UK. The figure bottomed out at just 1.86% in 2015, with the value for 2016 at 14.53% (Fig. 9.6). The hypothesis of a fall in the UK's attractiveness relative to other possible destinations of FDI is supported when examining the data on the inward FDI stock expressed as a share of the world's total. In 2016, the FDI stock in the

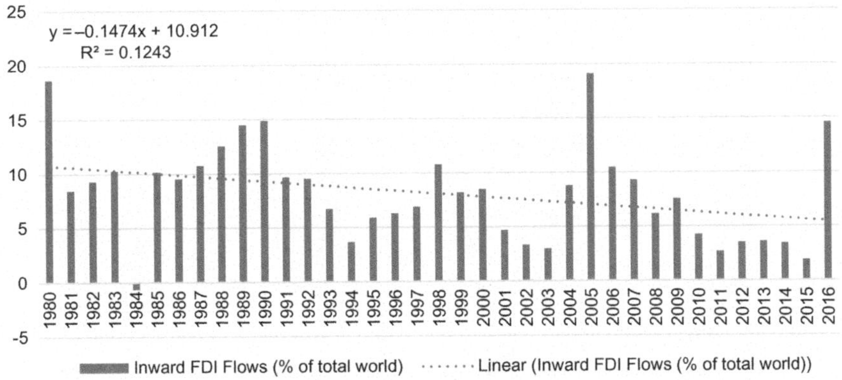

Fig. 9.6 Inward FDI flows in the UK as a percentage of world total (*Source* Author's own based on data from UNCTAD [2017b])

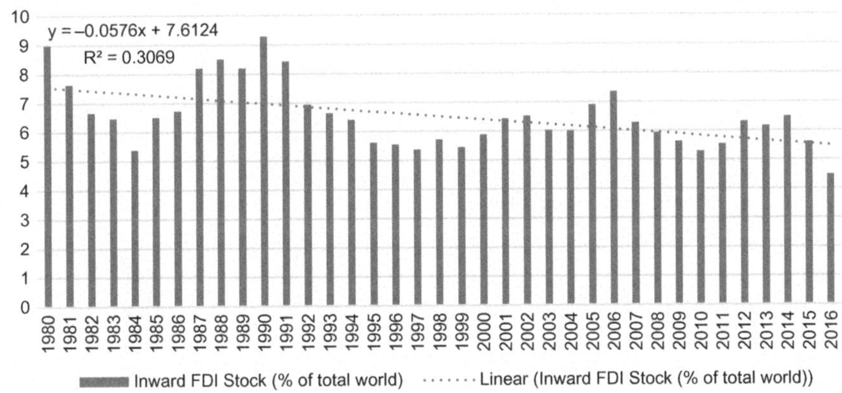

Fig. 9.7 Inward FDI stock in the UK as a percentage of world total (*Source* Author's own based on data from UNCTAD [2017b])

UK accounted for 4.47% of the world's total, about 2 percentage points less than the most recent maximum of 6.48% seen in 2014 (Fig. 9.7).

Analysing the sources of FDI International investment positions in the UK for the 2012–2015 period (Fig. 9.8), it can be observed that close to half of the inward FDI stock in the UK comes from EU members (45.37% in 2015). Therefore, the UK is heavily dependent on the EU for its inward FDI activity. This is further supported considering

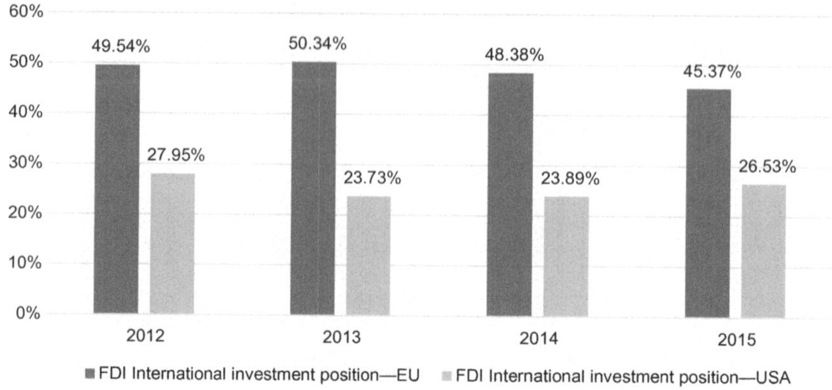

Fig. 9.8 FDI International investment positions in the UK by source as a percentage of total (*Source* Author's own based on data from Office of National Statistics [2016])

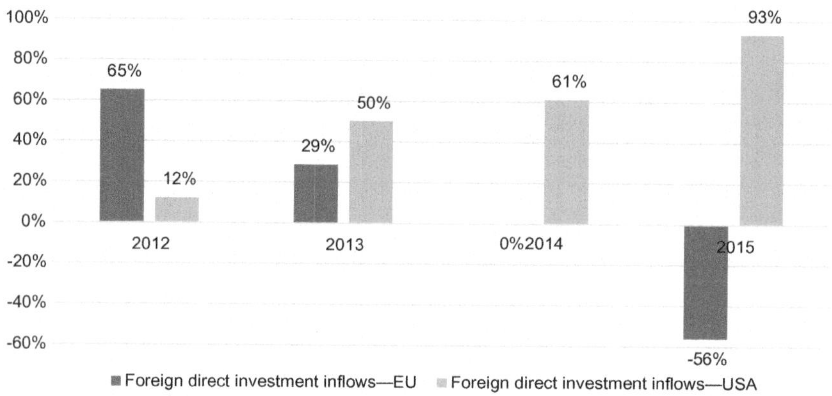

Fig. 9.9 FDI International investment flows in the UK by source as a percentage of total (*Source* Author's own based on data from Office of National Statistics [2016])

that the second-largest source of the FDI stock, the USA, accounts for about a quarter of the measured activity (26.53% in 2015). The importance of the EU as a source of FDI activity somewhat diminishes when examining flows (Fig. 9.9), while the importance of the USA increases. Interestingly, since 2014 inflows of FDI to the UK from the EU have been negative.

This data analysis has shown that even though the UK is a significant world destination for inward FDI, its relative attractiveness has been declining. Considering that about half of the FDI stock in the UK comes from the EU, it is justified to state that the UK leaving the EU will have a significant impact on inward FDI in the UK and therefore its economy.[8] However, as stated earlier, the UK's departure can have a positive or negative impact on inward FDI in the UK. To examine the hypothesised shock, the following empirics are presented.

9.4 The Relationship Between the UK's Trade Openness and Inward FDI Activity Within Britain

The aim of the empirical part of this chapter is to establish the cause-and-effect relationship between the UK's trade openness and inward FDI activity as well as to establish the causality, strength and direction of this relationship. This will make it possible to solve the issue raised at the end of the literature review, i.e. that an increase in trade openness can be associated with both higher and lower FDI. If the sign of the relationship is positive, then the UK's exit from the EU, by hindering trade, can be expected to lead to a fall in inward FDI in the UK and vice versa. Data for this study has been collected from the World Bank (2017) and UNCTAD (2017b), and covers the period from 1980 to 2016.

The first tool used in the study is the Granger causality test (Granger 1969) with a null hypothesis that trade openness does not Granger-cause inward FDI and a parallel opposite for the feedback hypothesis. Trade openness is measured as a sum of exports and imports expressed

[8]This conclusion is further supported by the works of Buch et al. (2001, 2003), Napiórkowski (2014), and Dhingra et al. (2016), which show that EU membership can be a significant determinant of inward FDI. In addition, Dhingra et al. (2016) show that "[s]triking a comprehensive trade deal—for example, joining Switzerland in the European Free Trade Association—would not significantly reduce the negative effects of Brexit on FDI". Interestingly, Simionescu (2017) suggests that the "UK should follow the model of Norway and Iceland after Brexit in order to avoid significant losses in the FDI inflows".

Table 9.1 Results of augmented Dickey–Fuller test for trade openness at levels

Null Hypothesis: TRADE_OPEN has a unit root			
Exogenous: constant			
Lag Length: 0 (Automatic—based on SIC, maxlag = 9)			
		t-Statistic	Prob.*
Augmented Dickey–Fuller test statistic		−1.319893	0.6101
Test critical values	1% level	−3.621023	
	5% level	−2.943427	
	10% level	−2.610263	

*One-sided *p*-values
Source Author's own based on data from the World Bank (2017)

Table 9.2 Results of augmented Dickey–Fuller test for trade openness at 1st differences

Null Hypothesis: D(TRADE_OPEN) has a unit root			
Exogenous: constant			
Lag Length: 0 (Automatic—based on SIC, maxlag = 9)			
		t-Statistic	Prob.*
Augmented Dickey–Fuller test statistic		−5.888136	0.0000
Test critical values	1% level	−3.621023	
	5% level	−2.943427	
	10% level	−2.610263	

*One-sided *p*-values
Source Author's own based on data from the World Bank (2017)

as a fraction of GDP, while inward FDI is expressed as its stock.[9] The stationarity of the variables required for the Granger causality test has been established with a set of Augmented Dickey–Fuller tests and the resulting differencing (for trade openness: Tables 9.1 and 9.2, and for the inward FDI stock: Tables 9.3, 9.4 and 9.5). The results of the test for two-period lags (Table 9.6) show that there is a causal relationship from trade openness to the inward FDI stock, which holds for three lags at a statistical significance level of 5% (Table 9.7). It also holds for four lags at a statistical significance level of 10% (Table 9.8), but becomes statistically insignificant for five lags (Table 9.9).

[9]The reason for excluding flows is the 2016 value, which significantly distorts the data.

Table 9.3 Results of augmented Dickey–Fuller test for inward FDI stock at levels

Null Hypothesis: I_FDI_S_UK_USD has a unit root			
Exogenous: constant			
Lag Length: 8 (Automatic—based on SIC, maxlag = 9)			
		t-Statistic	Prob.*
Augmented Dickey–Fuller test statistic		−1.110860	0.6971
Test critical values	1% level	−3.689194	
	5% level	−2.971853	
	10% level	−2.625121	

*One-sided *p*-values
Source Author's own based on data from UNCTAD (2017b)

Table 9.4 Results of augmented Dickey–Fuller test for inward FDI stock at 1st differences

Null Hypothesis: D(I_FDI_S_UK_USD) has a unit root			
Exogenous: constant			
Lag Length: 6 (Automatic—based on SIC, maxlag = 9)			
		t-Statistic	Prob.*
Augmented Dickey–Fuller test statistic		−1.218736	0.6524
Test critical values	1% level	−3.679322	
	5% level	−2.967767	
	10% level	−2.622989	

*One-sided *p*-values
Source Author's own based on data from UNCTAD (2017b)

Table 9.5 Results of augmented Dickey–Fuller test for inward FDI stock at 2nd differences

Null Hypothesis: D(I_FDI_S_UK_USD,2) has a unit root			
Exogenous: constant			
Lag Length: 5 (Automatic—based on SIC, maxlag = 9)			
		t-Statistic	Prob.*
Augmented Dickey–Fuller test statistic		−6.660523	0.0000
Test critical values	1% level	−3.679322	
	5% level	−2.967767	
	10% level	−2.622989	

*One-sided *p*-values
Source Author's own based on data from UNCTAD (2017b)

Table 9.6 Results of Granger causality test with two lags

Pairwise Granger causality tests			
Sample: 1980–2016			
Lags: 2			
Null Hypothesis	Obs.	*F*-Statistic	Prob.
D(I_FDI_S_UK_USD, 2) does not Granger-cause D(TRADE_OPEN)	33	0.36232	0.6993
D(TRADE_OPEN) does not Granger-cause D(I_FDI_S_UK_USD, 2)		7.00740	0.0034

Source Author's own based on data from the World Bank (2017) and UNCTAD (2017b)

Table 9.7 Results of Granger causality test with three lags

Pairwise Granger causality tests			
Sample: 1980–2016			
Lags: 3			
Null Hypothesis	Obs.	*F*-Statistic	Prob.
D(I_FDI_S_UK_USD, 2) does not Granger-cause D(TRADE_OPEN)	32	1.26328	0.3084
D(TRADE_OPEN) does not Granger-cause D(I_FDI_S_UK_USD, 2)		4.25865	0.0147

Source Author's own based on data from the World Bank (2017) and UNCTAD (2017b)

Table 9.8 Results of Granger causality test with four lags

Pairwise Granger causality tests			
Sample: 1980–2016			
Lags: 4			
Null Hypothesis	Obs.	*F*-Statistic	Prob.
D(I_FDI_S_UK_USD, 2) does not Granger-cause D(TRADE_OPEN)	31	0.76151	0.5615
D(TRADE_OPEN) does not Granger-cause D(I_FDI_S_UK_USD, 2)		2.48229	0.0735

Source Author's own based on data from the World Bank (2017) and UNCTAD (2017b)

Table 9.9 Results of Granger causality test with five lags

Pairwise Granger Causality Tests			
Sample: 1980–2016			
Lags: 5			
Null Hypothesis	Obs.	*F*-Statistic	Prob.
D(I_FDI_S_UK_USD, 2) does not Granger-cause D(TRADE_OPEN)	30	0.88671	0.5092
D(TRADE_OPEN) does not Granger-cause D(I_FDI_S_UK_USD, 2)		1.78970	0.1632

Source Author's own based on data from the World Bank (2017) and UNCTAD (2017b)

Table 9.10 Pearson correlation analysis

Covariance analysis: Ordinary		
Sample (adjusted): 1980–2016		
Included observations: 37 after adjustments		
Correlation *t*-Statistic Probability	TRADE_OPEN	I_FDI_S_UK_USD
TRADE_OPEN	1.000000 – –	
I_FDI_S_UK_USD	0.829289 8.779466 0.0000	1.000000 – –

Source Author's own based on data from the World Bank (2017) and UNCTAD (2017b)

The results of the Granger tests show that the trade openness of the UK is a determining factor for the inward FDI stock. Therefore, it is reasonable to expect that a change in trade openness will cause a change in inward FDI stock activity, which (as noted earlier) will see its reflection in the overall economy of the UK.

Pearson linear correlation coefficient (r) will be used to test the strength and sign of the established relationship. And so, the results (Table 9.10) show that there is a strong ($r = 0.829$), positive ($r > 0$) and statistically significant (p-value $= 0.000$; H_0: $r = 0$ is rejected at a 1% level of statistical significance) correlation between the UK's trade openness and the inward FDI stock in the UK.

The results of the correlation analysis show that high values of trade openness correspond to high values of the inward FDI stock and vice versa. Therefore, incorporating the results obtained from Granger analysis, it can be stated that with a fall in the UK's trade openness, the inward FDI stock in the UK will also decrease, as will the benefits from hosting FDI in the country.

9.5 Conclusions

The aim of this study was to establish the possible consequences of the UK leaving the EU from the perspective of foreign direct investment.

The literature was examined in order to (i) establish the benefits associated with hosting FDI (the scale of benefits would decrease with a fall in FDI activity in the UK) and (ii) to establish the importance of trade openness (which would decrease after the UK's departure from the EU) as one of the chief determinants of inward FDI. Next, with the use of Granger causality tests and the examination of Pearson's linear correlation coefficient, the direction, strength and sign of the relationship between trade openness and inward FDI were measured and established.

The results of the study show that, given the large number of areas affected by FDI in the host economy, the UK leaving the EU would—because of a fall in trade openness (and a fall in GDP) translating into a fall in inward FDI—have a negative impact on the UK's economy and that this impact will be dispersed throughout the economy.

These theory-based results fall in line with what has been concluded by other researchers. Welfens and Baier (2018) conclude that in "the hard Brexit case, the UK would lose almost half their FDI inflows from other European countries in the long run (20 years plus) …" (Welfens and Baier 2018, p. 18). Meanwhile, Dhingra et al. (2018), similar to Welfens and Baier (2018), find that the importance of the Brexit negotiations as the exit scenario will determine the magnitude of the studied impact as all of the possible post-Brexit models for the UK have significant disadvantages. The dependence on the exit scenario has also been highlighted by McGrattan and Waddle (2017), who argue "that the impact on investment, production, and welfare depends importantly on whether the United Kingdom acts unilaterally to block EU FDI or jointly with EU nations to erect cross-border barriers on each other's FDI" (McGrattan and Waddle 2017, p. 19). As for the estimate of the impact, it will depend not only on the exit scenario, but also on the empirical approach used. For example, Erken et al. (2018) "find much larger negative effects than most existing studies that use macroeconometric modelling to assess the effects of Brexit" (Erken et al. 2018, p. 46).

References

Ahmed, K. T., Ghani, G. M., Mohamad, N., & Derus, A. M. (2015). Does Inward FDI Crowd-Out Domestic Investment? Evidence from Uganda. *Procedia—Social and Behavioral Sciences, 172,* 419–426.

Azam, M., & Ahmed, A. M. (2015). Role of Human Capital and Foreign Direct Investment in Promoting Economic Growth. *International Journal of Social Economics, 42,* 98–111.

Barros, C. P., Caporale, G. M., & Damásio, B. (2013). *Foreign Direct Investment in the Asian Economies* (Economics and Finance Working Paper Series No. 13–20). Brunel University London.

Beer, F. M., & Cory, S. N. (1996). The Locational Determinants of U.S. Foreign Direct Investment in the European Union. *Journal of Financial and Strategic Decisions, 9*(2), 43–53.

Bevan, A. A., & Estrin, S. (2000). *The Determinants of Foreign Direct Investment in Transition Economies* (William Davidson Institute Working Paper No. 342).

Blonigen, B. A., & Piger, J. (2011). *Determinants of Foreign Direct Investment* (NBER Working Paper No. 16704).

Branstetter, L. (2006). Is Foreign Direct Investment a Channel of Knowledge Spillovers? Evidence from Japan's FDI in the United States. *Journal of International Economics, 68,* 325–344.

Buch, C. M., Kokta, R. M., & Piazolo, D. (2001). *Does the East Get What Would Otherwise Flow to the South? FDI Diversion in Europe* (Kiel Institute of World Economics Working Paper No. 1061).

Buch, C. M., Kleinert, J., & Toubal, F. (2003). *Where Enterprises Lead, People Follow? Links Between Migration and German FDI* (Kiel Institute for World Economics Working Paper No. 1190).

Carstensen, K., & Toubal, F. (2003). *Foreign Direct Investment in Central and Eastern European Countries: A Dynamic Panel Analysis* (Kiel Institute for World Economics Working Paper No. 1143).

Dhingra, S., Ottaviano, G., Sampson, T., & Van Reene, J. (2016). The Impact of Brexit on Foreign Investment in the UK. *BREXIT 2016 Policy Analysis from the Centre for Economic Performance* (pp. 24–33). London: Centre for Economic Performance, London School of Economics and Political Science.

Dhingra, S., Ottaviano, G., Rappoport, V., Sampson, T., & Thomas, C. (2018). UK Trade and FDI: A Post-Brexit Perspective. *Regional Science, 97*(1), 9–24.

Erken, H., Hayat, R., Prins, C., Heijmerikx, M., & de Vreede, I. (2018). Measuring the Permanent Costs of Brexit. *National Institute Economic Review, 244*.

Farla, K., De Crombrugghe, D., & Varspagen, B. (2016). Institutions, Foreign Direct Investment, and Domestic Investment: Crowding Out or Crowding In? *World Development, 88,* 1–9.

Granger, C. W. J. (1969). Investigating Causal Relations by Econometric Models and Cross-Spectral Methods. *Econometrica, 6*(37), 424–438.

Javorcik, B. S. (2015). Does FDI Bring Good Jobs to Host Countries? *World Bank Research Observer, 30,* 74–94.

Kalemli-Ozcan, S., & Nikolsko-Rzhevskyy, A. (2010). *Does Trade Cause Capital to Flow? Evidence from Historical Rainfalls* (National Bureau of Economic Research Working Paper No. 16034).

Kerr, I. A., & Peter, V. M. (2001). *The Determinants of Foreign Direct Investment in China*. Paper Prepared for Presentation at the 30th Annual Conference of Economics, University of Western Australia, Perth, September 23–26.

Leitão, N. C. (2010). Foreign Direct Investment: The Canadian Experience. *International Journal of Economics and Finance, 2*(4), 82–88.

Lipsey, R. E. (2002). *Home and Host Country Effects of FDI* (National Bureau of Economic Research Working Paper No. 9293).

Liu, W. S., Agbola, F. W., & Dzator, J. A. (2016). The Impact of FDI Spillovers Effects on Total Factor Productivity in the Chinese Electronic Industry: A Panel Data Analysis. *Journal of the Asia Pacific Economy, 21,* 217–234.

McGrattan, E. R., & Waddle, A. (2017). *The Impact of Brexit on Foreign Investment and Production* (Federal Reserve Bank of Minneapolis Research Department Staff Report No. 542).

Morrissey, O., & Udomkerdmongkol, M. (2012). Governance, Private Investment and Foreign Direct Investment in Developing Countries. *World Development, 40*(3), 437–445.

Morrissey, O., & Udomkerdmongkol, M. (2016). Response to 'Institutions, Foreign Direct Investment, and Domestic Investment: Crowding Out or Crowding In?' *World Development, 88,* 10–11.

Napiórkowski, T. M. (2013). Poland's Investment Attractiveness. In M. A. Weresa (Ed.), *Poland Competitiveness Report 2013: National and Regional Dimensions* (pp. 113–126). Warsaw: Warsaw School of Economics Press.

Napiórkowski, T. M. (2014). The Impact of Poland's Accession to the European Union on the Country's Foreign Direct Investment. In M. A. Weresa (Ed.), *Poland Competitiveness Report 2014: A Decade in the European Union* (pp. 113–133). Warsaw: Warsaw School of Economics Press.

Napiórkowski, T. M. (2015). Changes in Poland's Attractiveness to Foreign Investors in the Context of the Global Crisis. In M. A. Weresa (Ed.), *Poland Competitiveness Report 2015: Innovation and Poland's Performance in 2007–2014* (pp. 113–133). Warsaw: Warsaw School of Economics Press.

Narula, R., & Wakelin, K. (1997). *Pattern and Determinants of US Foreign Direct Investment in Industrialised Countries.* Maastricht: MERIT, Maastricht University.

Navaretti, G. B., & Venables, A. J. (2006). *Multinational Firms in the World Economy*. Princeton and Oxford: Princeton University Press.

Nunnenkamp, P. (2002a). *To What Extent Can Foreign Direct Investment Help Achieve International Development Goals?* (Kiel Institute of World Economics Working Paper No. 1128).

Nunnenkamp, P. (2002b). *Determinants of FDI in Developing Countries: Has Globalization Changed the Rules of the Game?* (Kiel Institute of World Economics Working Paper, No. 1122).

OECD. (2000). *Main Determinants and Impacts of Foreign Direct Investment on China's Economy* (Working Paper of International Investment No. 2000/4).

Office of National Statistics. (2016). *Inward Foreign Direct Investment (FDI) Involving UK Companies, 2015 (Directional Principle)* [online], https://www.ons.gov.uk/economy/nationalaccounts/balanceofpayments/bulletins/foreigndirectinvestmentinvolvingukcompanies/2015. Accessed October 25, 2017.

Pilbeam, K., & Oboleviciute, N. (2012). Does Foreign Direct Investment Crowd In or Crowd Out Domestic Investment? Evidence from the European Union. *The Journal of Economic Asymmetries, 9*(1), 89–104.

Przybylska, K. (2001). *Determinanty zagranicznych inwestycji bezpośrednich w teorii ekonomiczne. Empiryczna weryfikacja czynników lokalizacji zagranicznych inwestycji bezpośrednich w Czechach, Polsce i na Węgrzech* (Zeszyty

Naukowe Sesja Specjalna: Monografie No. 144). Akademia Ekonomiczna w Krakowie.

Simionescu, M. (2017). *The Influence of Brexit on the Foreign Direct Investment Projects and Inflows in the United Kingdom* (GLO Discussion Paper No. 68).

Stawicka, M. K. (2013). Czynniki warunkujące wybór miejsca lokowania bezpośrednich inwestycji zagranicznych w Unii Europejskiej – studium komparatystyczne lat 2005 i 2011. *Zarządzanie i Finanse, 11*(1), 315–329.

Svedin, D., & Stage, J. (2016). Impact of Foreign Direct Investment on Efficiency in Swedish Manufacturing. *Springerplus, 5*.

Szkorupová, Z. (2015). *Relationship Between Foreign Direct Investment and Domestic Investment in Selected Countries of Central and Eastern Europe* (Procedia Economics and Finance No. 23), pp. 1017–1022.

Te Velde, D. W. (2006). *Foreign Direct Investment and Development: An Historical Perspective.* Background paper for: World Economic and Social Survey for 2006, Commissioned by UNCTAD.

Temiz, D., & Gökmen, A. (2014). FDI Inflow as an International Business Operation by MNCs and Economic Growth: An Empirical Study on Turkey. *International Business Review, 23*(1), 145–154.

Tomohara, A., & Takii, A. (2011). Does Globalization Benefit Developing Countries? Effects of FDI on Local Wages. *Journal of Policy Modeling, 33,* 511–521.

Tülüce, N. S., & Doğan, İ. (2014). The Impact of Foreign Direct Investments on SMEs' Development. *Procedia—Social and Behavioral Sciences, 150,* 107–115.

UNCTAD. (2015a). *FDI Flows* [online], http://unctad.org/en/Pages/DIAE/FDI%20Statistics/FDIFlows.aspx. Accessed November 2, 2015.

UNCTAD. (2015b). *FDI Stock* [online], http://unctad.org/en/Pages/DIAE/FDI%20Statistics/FDIStock.aspx. Accessed November 2, 2015.

UNCTAD. (2017a). *Foreign Direct Investment (FDI)* [online], http://unctad.org/en/Pages/DIAE/Foreign-Direct-Investment-(FDI).aspx. Accessed September 29, 2017.

UNCTAD. (2017b). UNCTADStat [online], http://unctadstat.unctad.org/wds/ReportFolders/reportFolders.aspx?sRF_ActivePath=P,5,27&sRF_Expanded=,P,5,27. Accessed October 15, 2017.

Walsh, J. P., & Yu, J. (2010). *Determinants of Foreign Direct Investment: A Sectoral and Institutional Approach* (IMF Working Paper No. WP/10/187).

Welfens, P. J. J., & Baier, F. J. (2018). BREXIT and Foreign Direct Investment: Key Issues and New Empirical Findings. *International Journal of Financial Studies, 6*(2), 1–21.

World Bank. (2017). *World Economic Development Indicators* [online], http://databank.worldbank.org/data/reports.aspx?source=world-development-indicators. Accessed October 15, 2017.

Xun, L., & Awokuse, T. O. (2005). *The Determinants of US Outgoing FDI in the Food-Processing Sector*. Paper Prepared for Presentation at the American Agricultural Economics Association Annual Meeting, Providence, Rhode Island, July 24–27.

10

Brexit and Britain's Relations with South Korea

Piotr Ostaszewski

10.1 Introduction

The European perspective on Brexit reveals a diversity of calculations and predictions. When analysing statistics, various experts argue that Britain's departure will have a limited impact on the European Union. Some are even prompted to see Brexit as a win-win situation for both Britain and the UK. One thing is certain: the Brexit vote did not shake the EU's foundations for long. It rather had a short-lived impact, though it changed the official rhetoric that whatever happened would be a British problem with all its consequences.

But the Brexit effect is not limited to a European reaction. Britain is one of the world's most important economies, and it has been carefully watched by other countries since the Brexit vote. While the actual form in which Brexit will be implemented is still open to speculation, one has

---

P. Ostaszewski (✉)
Collegium of World Economy,
SGH Warsaw School of Economics, Warsaw, Poland
e-mail: piotr.ostaszewski@msz.gov.pl; charlott1@wp.pl

A. M. Kowalski (ed.), *Brexit and the Consequences for International Competitiveness*, https://doi.org/10.1007/978-3-030-03245-6_10

to ask about how Britain's main economic partners will react to the process once it gets underway.

This chapter analyses and evaluates the prospects of British–South Korean economic relations after Brexit by comparing the two countries' bargaining positions. To some extent, the evaluation must be based on speculation because Brexit is a long process that is set to be finalised in 2019.

10.2 Asian Countries and Brexit: Short Overview

Brexit might be perceived in the Asia-Pacific region as a new opportunity, not necessarily in a negative way. For many Asian countries, the EU, with Britain as a member, was a key destination for direct investment, but, as Jacob Wood and Haejin Jang point out, exports from most Asian nations have decreased recently, and Brexit might be a turning point to redefine their economic relations with Britain.[1] For countries such as Singapore, Myanmar and Malaysia, Brexit may pave the way to new economic and diplomatic relations. For stronger economies such as China, Japan and South Korea, Britain may offer better conditions for direct investment once it is freed from legal obstacles within the framework of the EU. Britain might renegotiate its free-trade agreement (FTA) with those three East Asian countries.

Looking at Britain's trade with Asian countries, one has to say that the UK on its own is not a major destination. Trade relations are scant, aside from the fact that Asian exports have been falling in recent years.[2] Therefore, Brexit is unlikely to have a dramatic impact on the UK's trade with Asia. Gauri Khandekar rightly pointed out in 2013, long before the Brexit vote, that the *EU remains Asia's largest trading partner*

[1]Jacob Wood, Haejin Jang, *Brexit: The Economic and Political Implications for Asia*, Social Sciences, 17 April 2017, http://www.readcube.com/articles/10.3390/socsci6020041.

[2]Aédan Mordecai, Pradumna Bickram Rana, Phidel Marion G.Vineles, *Brexit and Its Aftermath; Impact and Policy Recommendations for Asia*, Rajaratnam School of International Studies, December 2016, p. 10.

but its position is fast eroding relative to competitors.[3] Some scholars are convinced that Britain will be better positioned in commercial relations with Asia after Brexit. Others argue that Britain's position will weaken and become more vulnerable after the country leaves the EU.

10.3 South Korea and the European Union

To better understand the UK's future position in relations with South Korea after Brexit, it is necessary to analyse bilateral ties since this Asian country in 2009 signed a FTA with the European Union. That agreement formally entered into force in 2011. In 2010, the UK and South Korea signed a strategic partnership agreement. The EU–South Korea FTA is one of the most important and comprehensive FTAs that the EU has ever negotiated with an Asian partner. It has been of vital importance ever since South Korea's non-tariff barriers (strict protectionist policy) took effect.[4] However, thanks to the FTA, the EU entered the South Korean market more vigorously.[5]

Let's first look at the results of the tariff reduction/elimination in EU–South Korea trade. The EU sectors that have benefited the most are: machinery and appliances (70% of costs saved), chemicals, textile exports (93%) and agricultural products.[6] Apart from that, South Korea's most powerful tool is its so-called protectionist barriers, which, in bilateral relations, are a major obstacle to getting onto the South

[3]Gauri Khandekar, *EU-Asia Trade; in Need of a Strategy*, January 2013, p. 1, http://fride.org/download/PB_13_EU_Asia_trade.pdf.

[4]Generally, this policy is summarised by a 2016 Polish Trade Office report seen by this author (unpublished document). South Korea's NTB policy is analysed in http://benmuse.typepad.com/koreaus_fta/2010/07/what-are-these-korean-nontariff-barriers-to-us-auto-imports.html. See also Ivan Decreus, Cris Millner, and Nicholas Peridy, "Some New Insights into the Effects of the EU–South Korea Free Trade Area, The Role of Non-tariff Barriers", *Journal of Economic Integration*, 25 (4), December 2010, pp. 783–817.

[5]http://www.koreaherald.com/view.php?ud=20150727001080, see also Heunchong Kim, *Korea and the European Union: A Changing Landscape*, in Richard Youngs (ed.), *A New Context for the EU-Korean Relations*, FRIDE and the Korean Foundation, 2013, pp. 25–33.

[6]Karel de Gucht, *The EU—Korea Free Trade Agreement in Practice*, European Commission to Trade, Brussels, 2011, p. 5.

Korean market. There are special domestic regulations that are difficult to overcome. A case in point is Polish beef imports, which were banned in 2011 and have meagre prospects of reappearing on the South Korean market. Only strong economies—such as America or Canada—have realistic chances of successfully overcoming South Korean domestic barriers.[7]

South Korea's powerful economy (11th largest in the world in 2017) is among the EU's 10 most important and strategic partners.[8] In 2016, South Korea was the eighth largest supplier and ninth largest export market for the EU. The EU has meanwhile consolidated its position as South Korea's second largest supplier and third largest export market.[9] The EU's foreign direct investment (FDI) stock in South Korea increased by 8% from 2014 to 2015 (the latest year for which data are available) to EUR 49.7 billion, accounting for over 20% of the country's total FDI stock. Meanwhile, South Korean investment in the EU grew 19% in the same period to EUR 20.9 billion.[10]

10.4 South Korea and Britain: Economic Relations Before Brexit

Bilateral economic relations date back to the time of South Korean dictator Col. Bak Chung Hye, whose reforms proved to be of vital importance to the country achieving the status of a well-developed and stable economy that was capable of establishing relations with the world's most powerful nations on the basis of equality. The first UK–South

[7]Under the FTA, all NTBs in the automobile, pharmaceutical and electronics sectors are to be prevented. Although the FTA agreement mentions only four sectors as having specific commitments with considerable practical relevance, i.e. electronics, motor vehicles, pharmaceutical products/devices and chemicals, South Korea extensively uses its domestic trade barriers in relations with other—mostly weaker—economies.

[8]European Union External Action, *The Republic of Korea and the EU*, 10 May 2016, file:///C:/Users/PIOTR/Desktop/korea%20brexit/The%20Republic%20of%20Korea%20and%20the%20EU%20-%20European%20External%20Action%20Service.html.

[9]Ibidem.

[10]Ibidem.

Korea agreement was signed in 1976.[11] In 1996, the two countries followed up with a double taxation convention to regulate commercial relations.[12]

Today, more than 150 British companies operate in South Korea. These include well-known brands such as Jaguar Land Rover, BA, Bentley, Burberry and Standard Chartered. According to the British Department of Economic Trade, the strengths of South Korea's economy include the following: (1) The country is a global leader in electronics, shipbuilding and the steel and automotive sectors; (2) 74% of South Koreans undertake postgraduate-level education, and 7% of the country's gross domestic product (GDP) is spent on education; (3) South Korea has the highest level of broadband penetration in the world, with speeds of 100 megabytes; (4) The country leads the world in 4G mobile usage, with plans to invest USD 1.7 billion in 5G by 2020.[13] At the same time, UK companies enjoy benefits from the South Korean market due to a combination of factors: (1) the EU–South Korea FTA is estimated to be worth over half a billion pounds to UK business annually; (2) South Korea is a designated high-growth market for UK exports, increasing year-on-year since 2009, according to some

---

[11] *Agreement between the government of the Republic of Korea and the government of the United Kingdom of Great Britain and Northern Ireland for the promotion and protection of investments*, http://investmentpolicyhub.unctad.org/Download/TreatyFile/1843.

[12] *UK-Korea double taxation convention*, signed on 25 October 1996, https://www.gov.uk/government/uploads/system/uploads/attachment_data/file/498384/korea-dtc_-_in_force.pdf. British–South Korean economic relations date back to the end of the Korean War. At first, there were diplomatic and missionary activities, which were later supplemented by an economic presence and technical assistance in the 1969–1980 period. As well as this type of assistance, British finance and technology from the private sector played a major role in at least two important areas of South Korea's recent development: shipbuilding and the automobile industry. Reviving an old tradition, British banks have established themselves in South Korea in large numbers and have recently begun to operate in Pusan as well as Seoul. Recent years have seen the establishment of a number of joint venture companies, a trend that seems likely to grow. Two-way trade, minimal in the 1960s, has expanded rapidly in the last 10 years, with the balance heavily in South Korea's favour. A British Chamber of Commerce was established in Seoul in 1982—see J. Hoare, *The Centenary of Korean-British Diplomatic Relations: Aspects of British Interest and Involvement in Korea 1600–1983*, article presented before the Royal Asiatic Society-Korea Branch on 9 November 1983 in commemoration of the Korean-British centennial.

[13] Department for International Trade, *Doing Business in South Korea, South Korean Trade and Export Guide*, 25 November 2015, https://www.gov.uk/government/publications/exporting-to-south-korea/exporting-to-south-korea.

estimates; (3) The South Korean public has a taste for British culture and respect for UK goods; and (4) South Korea offers close proximity to the wider Asia-Pacific region and markets.[14]

The total value of South Korean exports to Britain in 2015 was estimated at around USD 9.14 billion, up from USD 6.88 billion in 2014.[15] The key commodities traded are special-purpose ships (33%), cars (16%), passenger and cargo ships (5.5%) and aircraft parts (2.1%). The total value of UK exports to South Korea was USD 6.47 billion, with crude petroleum accounting for 23% of the total, cars representing 12%, "packed medicaments" responsible for 2.5%, and "hard liquor" for 2.8%.[16] While South Korean exports to the UK grew, British exports to South Korea decreased by more than USD 1 billion in the same period (in 2014, they stood at USD 7.18 billion). This trend is only natural because South Korea's protectionist policy and soaring trade surplus are major elements of the country's economic policy. Notably, South Korea, unlike other Asian countries, has a trade surplus with China.[17]

According to the British Department for International Trade, strategic South Korean sectors include aerospace, creative industries, ICT, automotive, consumer products, fashion, food and drink, education, energy, environment, financial and legal services, life sciences and sports infrastructure. All these have been listed as priority sectors for Britain to create new opportunities for its industry ahead of Brexit.[18]

---

[14]Ibidem.

[15]*What Does South Korea Export to the United Kingdom*, OEC 2015, http://atlas.media.mit.edu/en/visualize/tree_map/hs92/export/kor/gbr/show/2015/.

[16]Ibidem.

[17]https://www.japantimes.co.jp/opinion/2017/03/26/commentary/world-commentary/china-south-korea-trade-war-must-end/#.Wf1bJmjWxKA. See also Stephen Denney, *South Korea's Economic Dependence on China*, 4 September 2015, *The Diplomat*, https://thediplomat.com/2015/09/south-koreas-economic-dependence-on-china/.

[18]Department for International Trade, *Doing Business in South Korea, South Korean Trade and Export Guide*, 25 November 2015, https://www.gov.uk/government/publications/exporting-to-south-korea/exporting-to-south-korea.

10.5 How Brexit Might Affect British–South Korean Relations

The EU will remain the second largest global economy post-Brexit, and the UK will be world No. 5. Of course, it will not be as influential as the whole of the EU, which, together with China and the USA, accounts for more than 48% of the global economy and for 43% of global trade and services. Still, Britain on its own will command 2.4% of global GDP, 2.1% of global trade in goods and 4.3% of global trade in services.[19] Statistics are important, and Brexit does not necessarily mean that Britain's position is at risk globally. Bilateral economic, business and trade relations may simply have to find a new dimension. Britain's bargaining position in bilateral relations will not necessarily diminish. Brexit will not mean severing economic ties with the EU, and therefore, British Foreign Secretary, Boris Johnson, had a point when he said that there are lots of *states queuing up for trade deals with the UK*.[20]

Undoubtedly, Britain may negotiate new agreements with large partners such as the USA, India, Hong Kong, China and Japan. The UK's new FTA with South Korea could be based on the existing EU-Korea FTA, though with fewer constraints and some revisions to ensure greater benefits on both sides.

On the other hand, a new trade agreement could not be based on the idea of copying and pasting from the EU–South Korea FTA. As Simon Hix and Hae-Won Jun have pointed out, *the precise terms of the EU–South Korea FTA were the result of a delicate compromise between a very large and powerful economy, the EU, and an economy and trading power less than one-tenth of the size of the EU*.[21] However, this does not mean that the EU–South Korea FTA was easier to negotiate because of the disproportionate size of the economies. South Korea had to offer

---

[19]Quote from Simon Hix, Hae-Won Jun, Can Global Britain forge a better trade deal with South Korea?, http://blogs.lse.ac.uk/brexit/2017/02/07/can-global-britain-forge-a-better-trade-deal-with-south-korea-this-is-why-its-unlikely/.

[20]Boris Johnson statement, https://www.politico.eu/article/boris-johnson-countries-queuing-up-for-post-brexit-trade-deals/.

[21]Quote from Simon Hix, Hae-Won Jun, ibidem.

some concessions to the EU (27 economies), whereas negotiations between two powerful economies could be more difficult, with fewer concessions on the South Korean side. Is Seoul going to be reluctant to replicate the terms of the EU-Korea FTA? Once again, the problem is not the terms because the UK is aware of South Korea's NTBs. Of course, no one expects any such negotiations to move smoothly because everyone protects their own economy. But, from the British perspective, it might be more difficult for a single trading partner to apply pressure on South Korea over its protectionist policy—to lift barriers or make them more flexible.

10.5.1 British Strategic Expectations

Embracing such a new approach would be a demanding task. British Secretary of State for International Trade, Liam Fox, sees Britain as a *global hub for international commerce*, which implies Britain taking the position of a single commercial and trade centre.[22] The crucial question is what kind of deal Britain strikes with the EU and whether the country remains part of Europe's single market customs union, plus whether British trade deals will still be controlled by Brussels.[23] If it remains part of the customs union, the UK will have to keep international tariffs on goods and services in the same way as EU members. Meanwhile, Britain seems to be determined to keep its position as a leader in global free trade, which would have an impact on its relations with South Korea.

10.5.2 South Korea's Strategic Expectations

South Korea, alongside Australia, New Zealand, Norway, China and India, has declared its readiness to redefine its economic and

[22]South Korea becomes the seventh country to agree to strike new trade links with Britain after Brexit as Liam Fox vows to make the UK a hub of global commerce, http://www.dailymail.co.uk/news/article-4045148/South-Korea-SEVENTH-country-agree-strike-new-trade-links-Britain-Brexit-Liam-Fox-vows-make-UK-hub-global-commerce.html.

[23]Ibidem.

commercial relations with Britain. Initially, however, the South Korean government responded with reservations to Brexit. In July 2016, the South Korean finance minister at the time, Yo Il-ho, spoke about a *prolonged effect* of Brexit and recommended a "sit and watch" tactic to see how the situation develops when Britain starts negotiations with the EU. For Seoul, it remained beyond any discussion that Brexit marked a new opportunity, so just a couple of weeks later the South Korean government was ready to define its economic and commercial strategy towards the UK. Yo Il-ho said that *trade negotiations with Britain should not take long or be dragged out.*[24] No dramatic change should not be expected in South Korea's strategy under its new liberal administration of President Moon Jae-in. There are no signals of the country thinking of revising its strategic expectations towards Britain.

10.5.3 Possible Problems for British Firms on the South Korean Market

One of the biggest problems British investors may encounter problems in South Korea after Brexit has to do with the country's legal services market. The top 50 law firms worldwide are of either US or UK origin, and UK law firms are free to open offices in the country under the EU–South Korea FTA. Licensed British lawyers can operate on the South Korean market and set up joint venture companies with South Korean law firms. So far five European law firms have opened offices in South Korea and all of these are British.[25] Britain leaving the EU may cause serious problems for UK law firms and force them to renegotiate their contracts based on a bilateral trade agreement.[26]

---

[24]Now South Korea wants British free-trade deal as UK ministers struggle to cope with demand, http://www.express.co.uk/news/world/688053/South-Korea-post-brexit-UK-trade-deal-Government-Australia-Canada-Germany-New-Zealand.

[25]These are Clifford Chance, Allen & Overy, Herbert Smith Freehills, Stephenson Harwood, and Linklaters, see Jun Hae-won, *Brexit and challenges for Korea-UK trade relations*, in IFANS Focus, January–June, 2017, p. 83.

[26]Simon Hix, Hae-Won Jun, ibidem.

According to Kim Sang-kyum, a law professor at Dongkuk University, *South Korea's legal market is relatively small in size and so far, it has mostly been the U.S. law firms who found reason to enter it, so the impact of Brexit will be limited.*[27] While he failed to mention UK law firms, it is expected that Brexit will not cause these firms to leave South Korea because the country has proven to be a profitable market for them.[28]

Another important problem is posed by the rules of origin under the EU–South Korea FTA, which stipulates that in car production, for instance, non-EU parts should not exceed 45% of the ex-works price of a car.[29] Some argue that, after Brexit, cars manufactured in the UK could *suddenly be subject to an 8% tariff when exported to Korea … even if the UK and South Korea agree to continue to apply the current EU–South Korea agreement!*[30]

10.5.4 Brexit as an Incentive for South Korea to Look for Another Major EU Partner

It is highly probable that, for many South Korean companies active in Europe, Germany will become a key destination for investment once Britain leaves the EU. Surprisingly, statistics show that Germany has the second largest South Korean community in Europe, after France. Germany's Frankfurt is home to more than 7000 South Koreans. They have settled and set up companies there. The region hosts big South Korean automotive and electronics companies, such as Hyundai/KIA, LG Electronics and Samsung (not to mention companies cooperating with these chaebols), and it might attract more South Korean settlement and investment, according to some scholars. Its geographical

[27] Brexit to have limited impact on Korean legal market: experts, http://www.koreaherald.com/view.php?ud=20160626000249.

[28] Ibidem.

[29] http://eeas.europa.eu/archives/delegations/south_korea/documents/eu_south_korea/presenter_1-2_rules_of_origin__en.pdf.

[30] Simon Hix, Hae-Won Jun, ibidem.

location adds to Frankfurt's appeal to South Korean investors as a hub for South Korean corporations.[31]

South Korea is the ninth largest investor in Germany. On the other hand, good conditions are also offered by Poland, which currently has the largest number of South Korean investment projects. More than 170 South Korean companies have invested in Poland, which makes South Korea the largest investor in the Central European region. The number of South Korean investors in Poland exceeds that in Germany, and the total value of South Korean investment in Poland is around USD 1.4 billion.[32]

Country	Number of South Korean investors	Value of South Korean investment, USD million
Poland	171	1490
Slovakia	106	1219
Czech Republic	63	1178
Romania	33	540
Hungary	83	412
Ukraine	32	252
Bulgaria	23	210
Austria	51	114
Sweden	21	80
Serbia	2	12
Latvia	3	8
Slovenia	4	7
Finland	15	6
Lithuania	5	2
Croatia	6	1

Source Warsaw Trade and Investment Office, Seoul, 2017

Of note is the Polish government's plan to build a huge hub airport—complete with a railway hub and other infrastructure—in the centre of the country in a project that might attract South Korean

[31]KPMG Wirtschaftprufungsgesllschaft, Brexit: An Impact Analysis, How Brexit may affect South Korean companies established in the UK, 2017, p. 10.

[32]Warsaw Trade and Investment Office in Seoul, materials delivered in 2017, non-printed version, author: Donat Krzysztof Wiśniewski, head of the office.

investors and lead to a shift in business to Central Europe. Brexit could prompt South Korean companies to join the project.

10.6 Conclusions

Overall, looking at Brexit from the bilateral British–South Korean perspective, it might be stated that: (i) Britain's position in relations with South Korea will not necessarily be weakened, though British law firms and the car industry, for example, may be adversely affected; (ii) Brexit will not negatively impact Britain's position in FTA negotiations with South Korea because South Korea has expressed its readiness to start such talks; and (iii) some Korean companies might be tempted to look for new destinations to do business, and Germany and Poland could offer interesting options.

References

Agreement Between the Government of the Republic of Korea and the Government of the United Kingdom of Great Britain and Northern Ireland for the Promotion and Protection of Investments. http://investmentpolicyhub.unctad.org/Download/TreatyFile/1843.

Boris Johnson Statement. https://www.politico.eu/article/boris-johnson-countries-queuing-up-for-post-brexit-trade-deals/.

Decreus, I., Millner, C., & Peridy, N. (2010, December). Some New Insights into the Effects of the EU–South Korea Free Trade Area, the Role of Non-tariff Barriers. *Journal of Economic Integration, 25*(4), 783–817.

de Gucht, K. (2011). *The EU—Korea Free Trade Agreement in Practice*. Brussels: European Commission to Trade.

Department for International Trade. (2015, November 25). *Doing Business in South Korea, South Korean Trade and Export Guide*. https://www.gov.uk/government/publications/exporting-to-south-korea/exporting-to-south-korea.

European Union External Action. (2016, May 10). *The Republic of Korea and the EU*. file:///C:/Users/PIOTR/Desktop/korea%20brexit/The%20Republic%20of%20Korea%20and%20the%20EU%20-%20European%20External%20Action%20Service.html.

Hoare, J. (1983). *The Centenary of Korean-British Diplomatic Relations: Aspects of British Interest and Involvement in Korea 1600–1983*. Article Presented Before the Royal Asiatic Society-Korea Branch on 9 November 1983 in Commemoration of the Korean-British Centennial.

Hix, S., & Jun, H.-W. (2017). *Can Global Britain Forge a Better Trade Deal with South Korea?* http://blogs.lse.ac.uk/brexit/2017/02/07/can-global-britain-forge-a-better-trade-deal-with-south-korea-this-is-why-its-unlikely/.

https://www.japantimes.co.jp/opinion/2017/03/26/commentary/world-commentary/china-south-korea-trade-war-must-end/#.Wf1bJmjWxKA. See also Denney, S. (2015, September 4). South Korea's Economic Dependence on China. *The Diplomat*. https://thediplomat.com/2015/09/south-koreas-economic-dependence-on-china/.

http://www.dailymail.co.uk/news/article-4045148/South-Korea-SEVENTH-country-agree-strike-new-trade-links-Britain-Brexit-Liam-Fox-vows-make-UK-hub-global-commerce.html.

http://www.express.co.uk/news/world/688053/South-Korea-post-brexit-UK-trade-deal-Government-Australia-Canada-Germany-New-Zealand.

http://www.koreaherald.com/view.php?ud=20160626000249.

http://eeas.europa.eu/archives/delegations/south_korea/documents/eu_south_korea/presenter_1-2_rules_of_origin__en.pdf.

Jun, H.-W. (2017, January–June). *Brexit and Challenges for Korea-UK Trade Relations*. IFANS Focus.

Khandekar, G. (2013, January). *EU-Asia Trade; in Need of a Strategy* (p. 1). http://fride.org/download/PB_13_EU_Asia_trade.pdf.

Kim, H. (2013). Korea and the European Union: A Changing Landscape. In R. Youngs (Ed.), *A New Context for the EU-Korean Relations, FRIDE and the Korean Foundation* (pp. 25–33).

KPMG Wirtschaftprufungsgesllschaft. (2017). *Brexit: An Impact Analysis, How Brexit May Affect South Korean Companies Established in the United Kingdom* (p. 10).

Mordecai, A., Rana, P. B., & Vineles, P. M. G. (2016, December). *Brexit and Its Aftermath; Impact and Policy Recommendations for Asia*. Rajaratnam School of International Studies.

UK-Korea Double Taxation Convention, Signed on 25 October 1996. https://www.gov.uk/government/uploads/system/uploads/attachment_data/file/498384/korea-dtc_-_in_force.pdf.

Warsaw Trade and Investment Office in Seoul, Materials Delivered in 2017, Non-printed Version, Author: Donat Krzysztof Wiśniewski, Head of the Office.

What Does South Korea Export to the United Kingdom, OEC 2015. http://atlas.media.mit.edu/en/visualize/tree_map/hs92/export/kor/gbr/show/2015/.

Wood, J., & Jang, H. (2017, April 17). *Brexit: The Economic and Political Implications for Asia, Social Sciences*. http://www.readcube.com/articles/10.3390/socsci6020041.

11

Does Brexit Influence China's "One Belt One Road" Initiative?

Günter Heiduk

11.1 Introduction

During his state visit to the UK in October 2015, China's President Xi Jinping declared that "China hopes to see a prosperous Europe and a united EU, and hopes Britain, as an important member of the EU, can play an even more positive and constructive role in promoting the deepening development of China-EU ties" (quoted in Wye 2016). Almost exactly two years earlier, during a visit to Kazakhstan, he presented the idea of a China-Central Asia cooperation belt. Referring to the ancient Silk Road, he proposed building a new "Silk Road Economic Belt". Shortly afterwards, while in Indonesia, he extended this vision to include a "21st-Century Maritime Silk Road".

In the wake of these declarations, China "invited" countries in Southeast Asia, the Middle East, Africa and Europe to join the initiative. Following authorisation by the State Council, the official

G. Heiduk (✉)
Collegium of World Economy,
SGH Warsaw School of Economics, Warsaw, Poland
e-mail: guenter.heiduk@uni-due.de

A. M. Kowalski (ed.), *Brexit and the Consequences for International Competitiveness*, https://doi.org/10.1007/978-3-030-03245-6_11

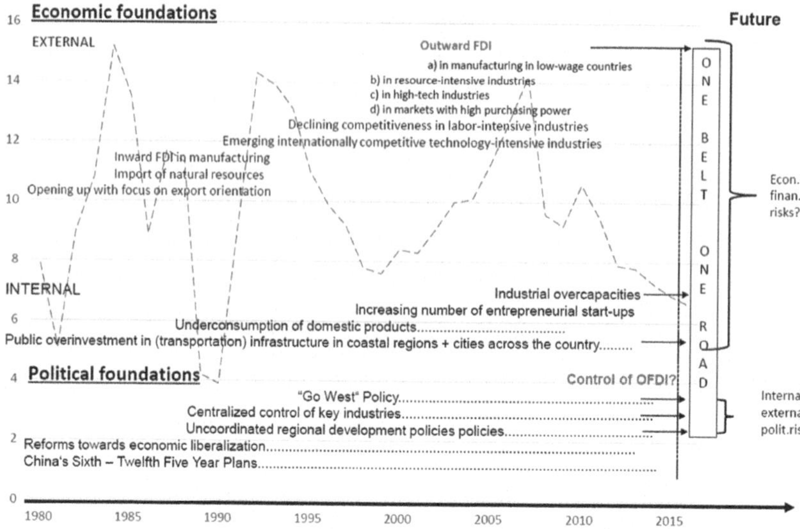

Fig. 11.1 Some characteristics of China's growth models (*Source* Own)

document on the "Vision and Actions on Jointly Building Silk Road Economic Belt and 21st-Century Maritime Silk Road" (NDRC 2015) was jointly issued by the National Development and Reform Commission (NDRC) and the involved ministries. The project is commonly known as "One Belt, One Road Initiative" (BRI).[1] From a domestic perspective, it aims to contribute to accelerating the development of China's western regions, serve as an outlet for overcapacities, notably in the steel industry, and improve China's external connectivity to Southeast, South and Central Asia, Europe and even Africa. It comes as part of China's long-term development strategy to become a new world power (Fig. 11.1).

After attracting inward foreign direct investment (IFDI) from Western countries and exporting low-value products in the 1980s and 1990s, from the 2000s onward China began promoting outward foreign direct investment (OFDI) in an effort to serve the growing energy

[1]The Chinese term *yidai yilu* literally means "one belt one road". The term "Belt and Road Initiative" (BRI or even B&R) is being increasingly used instead of "OBOR".

demand, strengthen presence in foreign markets and acquire foreign technology. In quantitative terms, China's OFDI promotion policy in 2016 lifted the country to the second largest home country for OFDI, with a 44% increase in such investment (UNCTAD 2017, p. 14). From China's government perspective, however, a growing part of the OFDI is not in line with the country's objectives. Following a warning at the end of 2016 by the State Administration of Foreign Exchange that it would "closely monitor irrational investment in real estate, hospitality, movie theatres, entertainment, and sports clubs" (SCMP 2017), the State Council's new guidelines on overseas investment classify outflows into the "encouraged, restricted and banned" categories (Table 11.1). This leads to the merger of the former OFDI promotion policy with the promotion of BRI investments: "According to the document, China will support eligible domestic enterprises to make overseas investment and join in the construction of projects in the Belt and Road Initiative. These enterprises should take the lead to export China's superior technology and equipment, upgrade the nation's research and manufacturing ability, and make up the shortage of energy and resources through prudent cooperation in oil, gas and other resources" (State Council 2017). In December 2017, the NDRC published new guidelines for OFDI by private companies (Covington and Burling 2017a) that

Table 11.1 Guidelines on overseas investment

Encouraged	Restricted	Banned
Infrastructure projects that facilitate the BRI	Investment in war zones or states without diplomatic relations with China	China's core military technology or products
China's advantaged equipment	Property, hotels, cinemas, entertainment and sports clubs	Technology or products banned from exports
High-tech and innovative research	Equity investment funds	Gambling and sex industries
Energy and mining exploration	Obsolete equipment	Investment that violates international treaties joined by China
Agriculture	Investment that violates environmental, energy or safety standards	Investment that hurts China's national interests or security
Commerce, culture, logistics and other sectors in the service industry		

Source State Council, People's Republic of China, Press Release, August 18, 2017

complement those released in August 2017 (Table 11.1). The new code of conduct addresses possible negative side effects of private OFDI, such as irregular operations; neglect of quality and safety; violations of local environmental protection laws; and neglect of strict risk controls. The NDRC announced the release of a similar code of conduct for state-owned enterprises (SOEs). The document is part of the government's efforts to cool the frenzy of overseas acquisitions and curb illicit capital outflows.[2]

Unlike China's opening-up strategies of the late 1970s (Fig. 11.1), which were clearly structured and gradually implemented in terms of content and time, the latest high-speed going-out phase topped by the BRI looks like a vision where incalculable economic, financial and even political risks are generously accepted. The official NDRC document focuses on the connectivity which BRI projects intend to create between Asia, Europe and Africa without implementing rigid structures in terms of geography, fields of activity, timing and budgets.

The vagueness of the BRI allowed the administration to move on with projects that were not primarily related to infrastructure improvements (Schubert 2017, p. 62). Moreover, the BRI triggered hectic gold-rush-type activities in various research institutes and universities, resulting in a huge number of books, movies, documentaries and exhibitions designed to revitalise the interest of the general public and win endorsement for historical Silk Roads and their modern-day relevance (Zhang 2016). Finally, it became clear that the scope of this initiative, from the perspective of the Chinese state, went beyond just infrastructure build-up (Zhang 2016). Putting the visionary idea into the current state of reality, a number of key cooperation projects—in fields such as infrastructure connectivity, industrial investment, resource development, economic and trade cooperation, financial cooperation, cultural exchanges, ecological protection and maritime cooperation—began to be promoted under the umbrella of the BRI. From a narrow economic point of view, it is hoped that this will make it possible to

[2]Published by *China Daily*, 18 December 2017, available at http://www.chinadaily.com.cn/a/201712/18/WS5a37841ea3108bc8c67356b2.html.

exploit new resources which can be traded by new transportation networks. This requires a closely coordinated concept of investment in resource exploitation, logistics and long-distance transportation networks (Mercator Institute 2018), an approach that was not completely successful so far. It is striking that four of the top five BRI projects are located in Asia. From a somewhat simplified view, BRI infrastructure projects are concentrated in Asia and Africa, with Chinese state-owned enterprises (SOEs) in transportation, construction and energy sectors as main contractors, whereas Chinese private companies are investing in Europe—including Central and Eastern European Countries (CEECs)—for market-, efficiency- and asset-seeking reasons. It can be assumed that an improved infrastructure along the Belt and Road routes will create incentives for Chinese private companies to expand their trade and foreign investment. From the Chinese government's perspective, the BRI is an important element in its future economic reform path to facilitate the catching-up development of China's western regions. But it goes without saying that the geographical and financial scope of the BRI enables the government to use it as a diplomatic instrument to increase China's geopolitical presence (Blanchard and Flint 2017; Rodriguez 2018). As overrepresented Chinese ownership in hard infrastructure may give rise for fears of foreign infiltration, the BRI contains the instrument of people-to-people dialogue as a core field of connectivity.

Despite an ongoing debate on the economic rationality of risky, large-scale infrastructure projects, the focus is not on developing and implementing efficient business and management strategies. Academics prefer to discuss whether the BRI is an instrument to pursue China's geopolitical and geoeconomic ambitions and to reduce its reliance on the US markets (Zhang 2016). Opinions range from consent to scepticism, even strict rejection. Zhang (2016) and numerous other scholars take an intermediate position by arguing that it needs more time to evaluate the effects of the BRI.

The intended geographical scope of the BRI underlines this evaluation. According to the official document (NDRC 2015), the BRI should connect regions and countries in Asia, Europe and Africa without describing basic principles which qualify for membership. However, the politics of the BRI points to China's main criteria, namely access to, first, foreign natural resources and, second, lucrative markets abroad. Therefore, the core areas are Southeast Asia, South Asia and Central Asia, while Eastern Europe, as a rather peripheral region, serves as a gateway to Western European markets. According to President Xi Jinping's intention, the BRI is open to all countries and even regional and international organisations such as the Shanghai Cooperation Organisation (SCO), the World Health Organisation (WHO), the United Nations Development Programme (UNDP) and the United Nations Industrial Development Organisation (OLGBIR 2017). Interest in joining the BRI may be signalled by countries which are members of the China-led Asian Infrastructure Investment Bank (AIIB), for example most EU member states in Western Europe. It should be noted that there seems to be a different view on the BRI between political leaders in Western Europe, on the one hand, and companies, on the other. British Prime Minister, Theresa May, refused to officially make the UK a member of the BRI programme, and German Chancellor, Angela Merkel, is demanding reciprocity in trade between Europe and China. According to a survey of the German Chambers of Commerce and Industry, almost 30% of German companies welcome the BRI whereas others regret the lack of information on the BRI (OBOReurope, 8 March 2018).

By the end of 2017, the Chinese government reported that 86 countries and international organisations have signed 100 cooperation agreements with China under the Belt and Road Initiative (Xinhua, 23 December 2017). The agreements cover a broad range of fields that include connectivity, production capacity, investment, economy and trade, finance, science and technology, society, humanities, quality of life and marine issues. The spatial shape of the Silk Road Economic Belt shows a tendency towards regionalisation, which is visualised by six economic corridors: the New Eurasian Land Bridge Economic Corridor, the China–Mongolia–Russia Economic Corridor, the China–Central Asia–West Asia Economic Corridor, the China–Indochina Peninsula Economic Corridor, the China–Pakistan Economic Corridor and the Bangladesh–China–India–Myanmar Economic Corridor.

Despite the fact that the BRI is a non-institutionalised "process in motion" so far (Kofner 2018), recent studies (e.g. Schubert 2017; Zhao 2016) trace the emerging discussions to establish a somehow institutionalised arrangement between the Silk Road Economic Belt and the Eurasian Economic Union (EAEU) or the Association of Southeast Asian Nations (ASEAN). It may come as a surprise that the BRI does not play a prominent role in China–EU relations. Critics point out that the BRI is open and flexible and that it deliberately avoids a unified institutional arrangement. It is obvious that China is transforming the *de jure* concept of openness into a de facto notion of "priority membership by invitation" for a number of "particular countries" which promise to deliver high benefits of infrastructure connectivity through railways, highways, sea transport, pipelines and/or information networks (OLGBRI 2017, pp. 10–11). With respect to China–UK relations, government officials in the UK claim that "Britain can be a natural partner in delivering infrastructure in Belt and Road countries by supporting the finance and planning needed".[3]

Freed from ideological ballast, Schubert (2017, p. 64) classifies the BRI as a "list of projects or things to do which have been almost emotionally connected to an idea about China and its increasingly important place in the world". In Tom Miller's words, "almost anything now can be counted as Belt and Road".[4] A large number of research-oriented studies on the BRI (e.g. PwC 2016) stress its effects on China's economic progress and the resulting changes of global powers.

Without going into details of the complex discussions and statements on the BRI, it can be concluded that the initiators and drivers of the BRI have yet to deliver convincing solutions for the purpose and impact of this project, its governance, institutional structure and its working modalities (Fues 2017, p. 3). Taking these challenges into account, the literature review provided in this chapter and suggestions on the impact of Brexit on China's BRI projects in the UK could only be a snapshot

[3]Quoted in *The Telegraph*, 29 May 2017, available at http://www.telegraph.co.uk/business/2017/05/29/britain-will-pay-key-role-building-chinas-new-silk-road/.

[4]Quoted by C. Campbell, in *Time* magazine, 12 May 2017, available at http://time.com/4776845/china-xi-jinping-belt-road-initiative-obor/.

based on insufficient, non-robust—if at all available—empirical data and partly contradictory statements from official and/or unofficial sources. Last but not least, the uncertainties of the timing, modalities and effects of Brexit considerably hamper an attempt to answer the main question posed in this chapter.

The chapter proceeds as follows: Sect. 11.2 highlights pre-Brexit China–UK economic relations with a focus on BRI projects, followed by post-Brexit scenarios in Sect. 11.3. A summary is presented in Sect. 11.4.

11.2 Pre-Brexit China–UK Economic Relations

The UK's country profile does not fit into the typical pattern of BRI members. In principle, the UK is neither an integral part of the Eurasian transportation network nor does it rely on Chinese infrastructure investment. Anyhow, the BRI has reached the UK, and it appears to play an increasing role in China–UK relations. Before describing the major BRI-related projects, a collection of official statements, followed by a brief overview of China–UK trade and investment relations, is intended to underline the importance which the Chinese and the British attribute to the BRI.

11.2.1 BRI in China–UK Relations

As already mentioned, the UK does not have the typical characteristics qualifying the country to be involved in the BRI. Summers (2016, p. 63) sees the UK is at best on the periphery of both the Silk Road Economic Belt and 21st Century Maritime Silk Road and points to the fact that many unofficial BRI maps terminate on the European mainland and do not include the UK. This might lead to the conclusion that the BRI does not play a significant role in the UK–China relationship. But in reality, there is a proactive response of government institutions, business organisations, universities, research institutes on the BRI. This also holds for the Chinese government. During his state visit to the

UK in October 2015, China's President Xi Jinping praised the bilateral relationship as a new "golden era". It is documented in the Sino-British Global Comprehensive Strategic Partnership. Prior to Xi's visit, the British finance minister visited China's Xinjiang province, thus demonstrating support for the BRI, which is designed to push forward the economic development of China's western provinces. The commitment of the British government to participate in BRI-related projects is also indicated by the UK's application to join the AIIB as its first European member. This engagement is in line with the economic interests of the British government to strengthen London's role as a financial mediator in BRI projects, especially when it comes to outside Chinese-financed investment in China's renminbi currency (RMB). Another special feature that distinguishes the UK's BRI involvement, compared with other European countries, is planned cooperation with China in third countries. Based on case studies, a report by the China–Britain Business Council (2016) shows that already existing projects along the Belt and Road demonstrate synergies between Chinese and UK firms. Even though official British statements on UK–China relations issued in 2016 did not make the BRI a priority,[5] the UK's engagement in the BRI is more open than that of other countries in the Western Hemisphere.

11.2.2 Trade in Goods and Services

China's access to the UK market is to a large extent shaped by EU rules and regulations. China's exports to the UK benefit from the economies of scale which the large EU28 market offers. The UK's exports to China compete with exports from other EU member states. For historical and language reasons, partly resulting from the British presence in Hong Kong, China has viewed the UK as a preferred gateway to the EU market. This applies not only to goods, but in particular services. In 2016, China ranked number three among the UK's largest import markets, accounting for 7% of the country's total imports. The top five product

[5]See, for example, Lord Keen of Elie's 24 November 2016 speech available at https://www.gov.uk/government/speeches/uk-china-legal-cooperation-along-the-belt-and-road.

categories were electronic equipment, machinery, furniture, toys/games and clothing. The first two product groups accounted for around 50% of the UK's total imports from China. China was the seventh largest export market for the UK—the second largest outside the EU—with a share of 4.4% on its total exports. This was far below Germany's exports to China (around 11%) and even below the Netherlands' exports (approximately 8%). However, since 2007, the UK has reported a strong growth in its exports to China. The top five exports from the UK to China were automobiles, machinery, precision instruments, oil/gas and pharmaceuticals. According to calculations, the UK is less vulnerable to a decline in Chinese imports than Germany.[6] The UK ranks ninth among China's largest trading partners, with a share of 2.7% in the country's total imports. Since 2014, UK–China trade has decreased slightly (Ward 2018, p. 5). In the past three years, China–Europe trade via freight trains on the three main routes of the Silk Road Economic Belt[7] has shown rapid growth (Kaplan 2016; Lobyrev et al. 2018). An up-to-date map of the train routes or reliable data on the frequency, volume and type of shipments are unavailable. Anyhow, they would become obsolete after a few days due to the dynamic growth of freight routes and frequency of freight trains. Furthermore, the final destination in Europe for a considerable number of trains is Germany's Duisburg, from where containers are distributed to other destinations. This also holds for the Yiwu–London freight train route. The main driving forces for China–Europe rail transport are the expansion of internationally integrated value chains, which considerably rely on just-in-time delivery, and relatively cheap transportation cost. It has to be noted that the dynamic growth of railway transportation from China to Europe is in part due to subsidies offered by Chinese authorities. According to a report by Janne Suokas (2018), these subsidies can be as high as USD 7000 per container, accounting for half of the total cost.

[6] *The Guardian*, "Xi Jinping's state visit: China's relationship with Britain in numbers", 20 October 2015, available at https://www.theguardian.com/news/datablog/2015/oct/20/xi-jinping-state-visit-chinas-relationship-with-britain-in-numbers.

[7] China–Mongolia–Russia Corridor, New Eurasian Land Bridge, China–Central Asia–West Asia Corridor (also called Trans-Caspian Rail Route).

Despite the dynamic growth, it can be assumed that the share of the combined trade volume on all China–EU land routes will be far lower than that of ship and air transportation. With currently available technologies, a limited number of products are suitable for land transport. Furthermore, the mainland routes already suffer from congestion and delays.

In contrast to trade in goods, the UK runs a surplus with China on trade in services. This surplus is particularly owed to the role of London as a top-tier financial centre. Chinese bankers expect that the BRI will cement London's position even after Brexit. They readily refer to London and New York as the global money centres and highly value London's expertise and the infrastructure.[8] British banks are already involved in the financing of BRI projects along the Silk Road Economic Belt. Evidence shows that growth in terms of projects and the regional expansion of the BRI bring financing to the fore. According to the Society for Worldwide Interbank Financial Telecommunication (SWIFT),[9] "London has retained its position as the preeminent foreign exchange (FX) and payment centre globally. It remains the dominant offshore hub for trading the Chinese renminbi (RMB) … London is the largest RMB payments centre outside of greater China with a 5.66% share".

11.2.3 FDI

From a global point of view, the UK is the number four host country for the IFDI stock (UNCTAD 2017). In Europe, it occupies the leading position (Hanemann and Huotari 2018, p. 32). There is no proven correlation between starting the BRI and regional changes of China's OFDI. Since China started its "Going Out" strategy, the UK has by far been the leading European destination for Chinese investors. They target a wide range of industries, including energy, transportation and

---

[8]Quoted in *The Telegraph*, 29 May 2017, available at http://www.telegraph.co.uk/business/2017/05/29/britain-will-pay-key-role-building-chinas-new-silk-road/.

[9]SWIFT, 25 April 2017, available at https://www.swift.com/news-events/press-releases/london-ranks-first-for-offshore-rmb-fx-transactions.

real estate, attracted by the country's stable regulatory and legal systems. Almost one-third of Chinese OFDI is in real estate in London and the wider region including Cambridge and Oxford.

The BRI era has been too short to verify whether it has changed China's OFDI in terms of regional direction and/or sectoral composition. However, statistics show that since 2016 China's OFDI in the EU has exceeded the volume of flows into the USA. Furthermore, the sectoral difference also points to a "BRI effect" in favour of Europe. While Chinese OFDI in the USA has been led by technology-intensive companies, in Europe, the energy and logistics sectors have received the bulk of Chinese investment (Casaburi 2016, p. 10). Cumulatively, approximately one-third of Chinese OFDI in the EU has targeted the energy and logistics/transportation sectors. In 2015, these sectors received almost 26% of China's total OFDI in the EU. The regional distribution of cumulative Chinese OFDI in the EU, which approached EUR 132 billion from 2000 to 2017 (Hanemann and Huotari 2018, p. 17), shows a clear lead of the UK. Meanwhile, statistics on the number of FDI projects originating from China (EY 2017, p. 24) show that Germany has attracted more Chinese investment projects than the UK since 2011. The EY report notes that the UK's FDI projects from both India and China are down, which the multinational professional services firm says is consistent with its earlier finding that the UK is not performing as well in attracting new projects as it is in securing FDI from existing investors. The broader geographic dispersion of Chinese OFDI across Europe is partly explainable with the Chinese interest to invest in infrastructure development and improvement in countries with a shortage of capital, which can also be somehow related to the BRI, according to EY. Last but not least, BRI-related investment projects may have contributed to the closing of a gap between EU FDI in China and Chinese FDI in Europe, the EY report says (Hanemann and Huotari 2018, p. 17).

Evidence suggests that the UK can keep its position as the number one recipient of FDI in Europe even after Brexit. According to an EY report (2017, p. 3), the post-Brexit UK will remain a successful destination for foreign investment. In 2016, London remained the most attractive European city with respect to FDI inflows, but its lead over

Paris had shrunk. Some forward-looking indicators in EY's survey suggest that future success is far from guaranteed. Nor have the British government's fluctuating statements and actions regarding Chinese FDI boosted the confidence of Chinese investors (see the "Hinkley case" below).

11.2.4 BRI Projects in UK

Yiwu–London Freight Train Route

Since 2013, when the first "Yixinou" freight train linking China's Chongqing and Germany's Duisburg arrived at its final destination, the China–Europe railway network has grown rapidly. In September 2017, China's Xinhua[10] news agency reported that China had 51 Sino–European freight train routes, with trains from 28 Chinese cities travelling to 29 cities in 11 European countries. In November 2017, a new train route from Dalian to Bratislava, Slovakia, was added to the network. One of the leading Chinese land ports of departure is Yiwu, a city near the central eastern coast of China. Dubbed the "world supermarket", Yiwu is the starting point for nine freight train routes to Europe. Referring to Chinese customs statistics, Xinhua (9 September 2017) reported that Yiwu's Sino–Europe freight service reached CNY 3 billion in 2016. The Yiwu–London train route opened with celebrations on 2 January 2017. Eighteen days later the train arrived at its destination of Barking, east London.

Forbes[11] heralded this new train route as a new stage in China–UK trade relations. More important than the relatively small volume of transported goods is the conceptual link between China and the UK when considering trade opportunities in Central Asia, Russia and post-Brexit mainland Europe. Household goods, clothes and shoes account

[10]Xinhuanet, 9 September 2017, available at http://news.xinhuanet.com/english/2017-09/09/c_136597151.htm.

[11]*Forbes*, "Trains are the new Pandas", 6 January 2017, available at https://www.forbes.com/sites/wadeshepard/2017/01/06/the-story-behind-the-new-china-to-uk-train/#8d676fb261b4.

for the bulk of products on the Yiwu–London train service. The train runs as a weekly service. A major weakness of the route is its low frequency, especially in the west–east direction. The first London–Yiwu train arrived at its destination on 29 April 2017, loaded with pharmaceuticals, baby products and alcoholic drinks. The demand for eastbound rail-shipped products is obviously lower than in the opposite direction.

It is noteworthy that not all the wagons/containers that leave Yiwu arrive in London. Duisburg serves as a distribution hub where containers are reloaded to trains bound for Madrid, Milan and London. This means that one train delivers products to EU member states with one customs document for the EU28. The customs clearance process (declaration and inspection) can be completed at once. ID labels, import tariffs, duty rates, trading standards and food safety requirements (with some exceptions) do not need to be differentiated according to destinations in different countries. Last but not least, the Yiwu–London connection can serve as an important component to establish distant China–UK production networks. There is some evidence that the UK is re-industrialising and Chinese investment and intra-industry trade in parts and components may play a considerable role. Due to the lack of detailed data on the volume, prices and type of products, it is impossible to analyse the interdependence between trade by rail and bilateral investment. Therefore, quantitative proof cannot be provided.

It is expected that the Chinese government will stop subsidising additional rail-freight capacity in 2020, which will slow network expansion.[12] The continuing long-term existence of the Yiwu–London line depends not so much on subsidies as the probable loss of competitiveness due to Brexit. Even if data on the trade volume for this route were available, the short time period would not make it possible to draw conclusions about the trade expansion effect. This would require

[12] *The Economist*, "New rail routes between China and Europe will change the trade pattern", 16 September 2017, available at https://www.economist.com/news/business/21728981-new-silk-railroad-will-challenge-airlines-and-shipping-firms-new-rail-routes-between-china.

information on the shift from sea transport to railway transport. The most important factor for the long-term success of the freight route is future arrangements between the EU and the UK as well as between the UK and China.

Infrastructure Investment: The Case of the Hinkley Point C Nuclear Power Station

BRI and Juncker Plan

Even though the destination of almost all new BRI freight rail routes is a city in the EU, the EU has voiced mixed signals to China. Le Corre (2017) points out to EU's reservations about China which were documented in the vote against China's application to grant the country the market economy status under WTO law. Further unsolved issues are the reciprocity clause and the access of European companies to the Chinese market. This rather negative view seems to be slowly changing. On the EU level, a report by the European Parliamentary Research Service (2016) states that China's infrastructure investment in Europe until recently targeted individual EU countries and the 16+1 group rather than the EU as a bloc. The recently established EU–China Connectivity Platform[13] aims to create synergies between the BRI and the Juncker Plan (the European Commission's Investment Plan for Europe) announced in 2014. At the EU–China High Level Dialogue, in September 2015 both sides reaffirmed their strong interest in each other's flagship initiatives, the BRI and the Juncker Plan. It is hardly surprising that the majority of the TEN-T-related projects which are currently on the agenda of the EU–China Connectivity Platform are almost exclusively located in CEECs.[14] This coincides with China's growing interest in special relations with CEECs, as evidenced by the

[13]See Article 6 of the 17th EU–China Summit Joint Statement, available at https://eeas.europa.eu/sites/eeas/files/eu-china_summit_2015_joint_statement.pdf.

[14]List of TEN-T-related projects presented in May 2017, available at https://ec.europa.eu/transport/sites/transport/files/ten-t-rel-projects-may-2017.pdf.

16+1 Cooperation Framework. This format serves as a platform for the implementation of BRI projects in the region. For the EU, it is an 11+5+1 framework because five countries do not belong to the EU. Anyhow, the European Parliament and the European Commission are aware that the BRI brings opportunities and challenges for the European transport system, which calls for intensive cooperation (Steer Davis Gleave 2018). A major concern from the European Union is the lack of a clear BRI investment plan.

Hinkley Point C Nuclear Power Plant

The slow pace of progress in implementing BRI and Juncker Plan-supported infrastructure projects at the EU level seems to contrast with the interest and actions in EU member states that are keen to preserve privileged bilateral ties to China. If the special relations of CEECs with China are ignored, the UK is the most prominent country that could be a "natural partner" for China's new Silk Road programme. The British finance minister said at the 2017 Belt and Road Forum: "As China drives forward the Belt and Road initiative from the east, we in Britain are a natural partner in the west, standing ready to work with all Belt and Road partner countries to make a success of this initiative".[15] It is realistic to anticipate that this statement targets the British post-BREXIT era where the UK needs new opportunities for growth-enhancing and sustainable external relations. Britain's efforts to put its relations with China on a firm basis became apparent already with the signing of the "China-UK Joint Declaration on Building a Global Comprehensive Strategic Partnership for the 21st Century" when President Xi Jinping paid a state visit to the UK in October 2015.[16] Paragraph 11 of the Declaration states that "[b]oth sides have a strong

[15]Quoted by Reuters, "Britain says it's a natural partner for China's new Silk Road", 14 May 2017, available at https://www.reuters.com/article/us-china-silkroad-britain/britain-says-its-a-natural-partner-for-chinas-new-silk-road-idUSKBN18A04D.

[16]The full text of the China–UK Joint Declaration is available at https://china.cgtnamerica.com/2015/10/22/full-text-of-china-uk-joint-declaration/.

interest in cooperating on each other's major initiatives, namely China's 'Belt and Road' initiative and the UK's National Infrastructure Plan and the Northern Powerhouse. They will further discuss a China-UK infrastructure alliance under existing mechanisms and explore cooperation in light of the China-EU Joint Investment Fund and Connectivity Platform. This includes supporting cooperation between China and the European Investment Bank".

China's involvement in the Hinkley Point C nuclear power station has a history going back to 2005.[17] The UK prime minister at the time, Tony Blair, launched an energy review that resulted in giving the formal go-ahead to initiate, fund, construct and operate nuclear plants by private companies. In 2008, French nuclear energy company EDF bought British Energy, which owned Britain's existing nuclear power plants. EDF unveiled plans to build four new nuclear reactors in the UK. After an intense debate on government subsidies, in 2013 EDF and the British government reached a provisional agreement. One year later, the European Commission approved a British state aid decision. Due to a bottleneck in financing the project, in October 2015 EDF signed a deal with China General Nuclear Power Corporation (CGN), which committed China's General Power Group (CGN) to cover 33% of the costs, estimated at GBP 18 billion. Furthermore, the deal guaranteed CGN a 20% stake in developing a second EDF-led plant at Sizewell and opened the door for developing a reactor in Bradwell with Chinese technology. This would mark the first wholly Chinese-designed nuclear reactor to be built in a Western country. Britain's then-Prime Minister David Cameron hailed the deal as "historic". In July 2016, the new British Prime Minister Theresa May felt compelled to suspend the Hinkley project while a security review was carried out. This was decided a day before the scheduled execution of the agreement between EDF, CGN and the British government and put the relationship between the British and Chinese governments under strain. Based on a new agreement regarding safety checks, the British government

[17] *The Telegraph*, 19 September 2016, available at http://www.telegraph.co.uk/business/0/hinkley-point-c-new-nuclear-plant-timeline-of-the-story-so-far/.

formally approved the construction of the plant on 15 September 2016. The Hinkley Point C nuclear power station should come on stream in 2025. European energy suppliers admitted that the deal was a "major coup for China's nuclear export effort which is part of its grand 'Belt & Road' effort",[18] whereas numerous British media outlets[19] voiced concerns about safety, security, energy prices and an influx of Chinese workers. From the Chinese perspective, the positive decision created the hope for Chinese companies to play a significant role in future reactors at Sizewell in Suffolk and Bradwell in Essex, which obviously increased Theresa May's concerns. Government approval for the latter project is still pending. In 2018, a new controversy was ignited by signs that China's CGN was interested in buying a major stake in eight power stations, including Sizewell in Suffolk and Dungeness in Kent.[20]

Infrastructure Investment: The Case of London's Royal Docks

In October 2017, Chinese developer Advanced Business Park (ABP) announced the official start of construction of a GBP 1.7 billion development at the Royal Albert Dock (ABP 2017, p. 3). The development forms an important part of the westernmost end of the Silk Road Economic Belt. London's next business district should serve as a new gateway to Europe for Asian companies. The location is close to the Yiwu-London final train station and the London City airport. The cluster concept is similar to that of the Chongqing international logistics hub for the BRI.[21] ABP has succeeded in attracting Chinese state-owned conglomerate CITIC as both the investor and lead contractor for the project.

[18] *Energy Post*, 1 December 2017, available at http://energypost.eu/looking-ahead-to-2018-for-nuclear-energy/.

[19] *The Guardian*, on 21 December 2017, noted that Hinkley Point C would cost twice as much as the 2012 Olympic Games, available at https://www.theguardian.com/news/2017/dec/21/hinkley-point-c-dreadful-deal-behind-worlds-most-expensive-power-plant.

[20] *The Guardian*, 8 July 2018, available at https://www.theguardian.com/environment/2018/jul/08/china-interested-majority-stake-uk-nuclear-power-stations-reports.

[21] Expat Club, 1 September 2017, available at http://www.cqexpat.com/i_forum/show.asp?id=67783.

Full completion is expected for 2026. It is worth noting that ABP has learned from bad experiences of Chinese building contractors in other large BRI projects, especially in Southeast Asia. The mass influx of Chinese workers led to heavy protests among local workers and trade unions. In order to avoid similar reactions in London, ABP has made a commitment to ensure that newly created jobs are accessible to local workers. But there is evidence that parts of the construction project will be carried out by Chinese companies. With the support of the Guangdong province government, ABP signed a memorandum of understanding (MoU) with Tian An Group to build a centre to accommodate companies from this province (ABP 2017, p. 5). Establishing subsidiaries or even European headquarters should boost the investors' exports to the UK and continental Europe. The Guangzhou municipal government has offered to work together with British organisations to attract British investment in the wider Guangzhou region. According to official statistics, trade between Guangzhou and the UK totalled RMB 16.38 billion in 2016, while Guangzhou investment in the UK reached USD 250 million (ABP 2017, p. 5). Needless to say that ABP's commitment to develop the district can be interpreted as a sign of confidence in the UK economy post-Brexit and its continually important position in Europe's economy. According to announcements from British and Chinese government officials, business association leaders and managers, the future of the new London Royal Dock business district is considered as decidedly positive. A lone voice from a Green Party member criticised that "[we] shouldn't be giving public land, tax breaks and other generous inducements to large corporations. The mayor can steward international investment for the common good, instead of flogging our city off to the highest bidder. I'd like to see the new mayor renegotiate the deal to ensure fair taxes, a decent amount of social housing and genuinely public space managed by the local authority".[22] In June 2018, Chinese developer ABP signed a memorandum of understanding with the ExCeL International Exhibition Centre to jointly attract international enterprises to the Royal Docks to drive economic

[22]Quoted in the Pulitzer Center, 7 June 2016, available at https://pulitzercenter.org/reporting/selling-silverware-how-londons-historic-dock-was-sold-chinese.

growth for east London.[23] According to ABP, the first phase of around 650,000 sq ft of development is due to open in early 2019.

London: Global Financial Centre

It is estimated that BRI countries require anywhere from USD 1 trillion to USD 5 trillion in financing to meet their needs for modern infrastructure, especially in transport and energy, over the next five years. This amount far exceeds China's financial capacity. China cannot rely on its own banks, including the China-backed AIIB, the Silk Road Fund and the NDB, to generate funding for the BRI. Garcia-Herrero (2017, p. 16) concludes that the key source of co-financing would logically be Europe. This inevitably leads to London as a top-tier global financial centre. The London Stock Exchange "is on hand to facilitate this initiative".[24]

As the number of BRI projects in many countries is rapidly growing, the demand for money to finance them is also increasing. The Belt and Road Forum in Beijing in May 2017 clearly showed that the political rhetoric of celebrating the BRI as a project that will change the world is increasingly shifting towards actions and ideas on how to finance them. In this respect, UK Finance Minister Philip Hammond never gets tired of pointing out that "Britain can be a natural partner in delivering infrastructure in Belt and Road countries by supporting the finance and planning needed".[25] The long-term successful functioning of BRI projects heavily depends on their sound financing as well as on accompanying consulting services. Chinese government officials are obviously convinced that London will play a pivotal role in raising finance. It is no surprise that British banks are highlighting London as

[23]Information provided by ABP, available at http://www.abp-london.co.uk/media/press-releases/london-royal-docks-organisations-join-forces-to-create-a-global-destination-for-innovation-and-technology.

[24]London Stock Exchange Group, Belt and Road Initiative, available at https://www.lseg.com/markets-products-and-services/our-markets/london-stock-exchange/belt-and-road-initiative-bri.

[25]*The Telegraph*, 29 May 2017, available at http://www.telegraph.co.uk/business/2017/05/29/britain-will-pay-key-role-building-chinas-new-silk-road/.

the number one location for BRI financing. A Standard Chartered Bank executive[26] has noted that no other city in the world matches London's expertise and financial infrastructure. It is obvious that New York and the Americans in general don't show great passion for BRI. British and Chinese financial markets will be getting closer, integrated by plans to create a Stock Connect electronic trading link between the London Stock Exchange and the Shanghai Stock Exchange, to enable investors to conduct cross-border trading. The CEO of the London Stock Exchange, Nikhil Rathi, has noted that the London Stock Exchange has played a major role in supporting companies in BRI countries to access international capital markets.[27]

More recently, the Chinese government has increased its efforts to make domestic and BRI-included infrastructure projects cleaner. The British and Chinese government have already started working together to set up joint international standards and studies on how best to manage green projects.[28] Officials from the City of London are keen on "promoting green finance in China by setting standards and platforms for companies to issue green bonds and other products, to raise funds to finance infrastructure projects that could help cut emissions".[29] London is widely recognised as the leader in green financing. According to the China Green Finance Committee, "BRI provides a timely platform for the timely delivery to help meet the estimated USD 22.6 trillion infrastructure funding gap required until 2030 in Asia and the Pacific".[30]

The London Stock Exchange also plays a leading role in the internationalisation of the renminbi. In October 2014, the British government issued the first non-Chinese sovereign offshore RMB bond. Less

[26]Sam XU, quoted in *The Telegraph*, 28 May 2017, available at http://www.telegraph.co.uk/business/2017/05/29/britain-will-pay-key-role-building-chinas-new-silk-road/.

[27]Nikhil Rathi, quoted in "The Belt and Road Initiative", London Stock Exchange Group, available at https://www.lseg.com/markets-products-and-services/our-markets/london-stock-exchange/belt-and-road-initiative-bri.

[28]Finance is classified as "green" if it aims to reduce carbon emissions or raise resource efficiency.

[29]The deputy chairman of the policy and resources committee for the City of London, quoted in the *South China Morning Post*, 7 May 2017, available at http://www.scmp.com/business/china-business/article/2093306/london-calling-green-finance-belt-and-road-latest-china.

[30]Ma Jun, quoted at the London Stock Exchange Group, available at https://www.lseg.com/markets-products-and-services/our-markets/london-stock-exchange/belt-and-road-initiative-bri.

than two years later, China's Ministry of Finance completed a CNY 3 billion bond listing on the London Stock Exchange, its first offshore RMB sovereign bond ever listed outside Greater China. Outside Hong Kong, London has become the largest market for renminbi-denominated debt. According to Bourse Consult (cited in Subacchi and Oxenford 2017, p. 7) and SWIFT (2018), London is the leading location in offshore foreign exchange trading for renminbi outside China. The latter reports that London's share in renminbi trading is around 26% of the total value traded, whereas in terms of volume London tops the list with around 41%. Liu et al. (2017a, p. 16) note that the ongoing development of London as a centre for renminbi trading could help to promote the BRI as well as to support the renminbi internationalisation in generally. Regarding project financing, London's Private Finance Initiative has been in place since 1992. London's hard and soft infrastructure provides a unique asset for financing BRI projects. Anyhow, Liu et al. (2017b, p. 1) point to the fact that there is still a lack of liquidity in offshore renminbi markets. They propose to issue renminbi-denominated debt financing instruments that are linked to BRI projects which requires cooperation between British and Chinese government and private banks. They recommend a number of policy measures to put the City of London in the central focus of financing activities that increase the internationalisation of the renminbi.

It is common knowledge that financial business significantly relies on soft factors such as trust, competence, knowledge which means that the City of London's role in the BRI goes beyond its role as pure financier. BRI-related UK–China business partnerships need the "City's" know-how in risk management as well as the London-based professional services in law and consulting.

11.3 Post-Brexit Scenarios

11.3.1 UK–China Relations Post-Brexit

China's exposure and reaction to Brexit will primarily depend on the exit modalities on which the UK and the EU agree. Furthermore, the functioning of the transmission channels with respect to trade and

FDI might create deviations from the expected effects of the model, which needs trial-and-error adjustments. The current state of negotiations between the UK and the EU suggests that, for different reasons, the Norwegian-style EEA agreement, the Swiss-style bilateral accord and the Turkish-style customs union are all unrealistic.[31] All models are caught in a trade-off dilemma between political costs and economic benefits. The negotiations so far indicate that, from the British point of view, the willingness to minimise political costs and maximise economic benefits is not an option. From the Chinese point of view, economic relations with the UK need to be negotiated and put on agreements on market access for goods, services and capital differing from the current EU conditions. The UK, as a small country compared to China (as well as the USA), would have rather limited bargaining power in trade and investment negotiations. As a result, China might negotiate more favourable market access conditions with the UK compared to the current conditions under the umbrella of the EU. But, this does not automatically result in more trade with and more FDI flows to the UK.

As the Belt and Road Forum opened in Beijing in May 2017, Britain's Finance Minister Philip Hammond emphasised Britain's role as a natural partner for China's new Silk Road programme. "Britain is keen to sign a free trade deal with China when it exits the EU, and while Beijing has repeatedly expressed support for closer European integration, it has also said that Britain's withdrawal from the bloc would not affect ties … As China drives forward the Belt and Road initiative from the east, we in Britain are a natural partner in the west, standing ready to work with all Belt and Road partner countries to make a success of this initiative … As we embark on a new chapter in our history, as we leave the European Union, we want to maintain a close and open trading partnership with our European neighbours, and at the same time pursue our ambition to secure free trade agreements around the

[31]The models are described and discussed, e.g., in the Global Counsel (2015), Dhinga and Sampson (2016), and Emerson (2016).

world with new partners and old allies alike".[32] Regardless of the result of the Brexit negotiations with the EU, the British government needs to quickly start negotiations on a free-trade agreement with China.

China's ambassador to the UK expects that Brexit will impact UK–China relations by offering new development opportunities.[33] He noted that "China and the UK, as the world's second- and fifth-largest economies, have much to offer to each other, for example through trade, investment, finance, science and technology, and many other areas. The untapped potential is huge … After the UK voted to leave the EU, many Chinese companies cast a 'vote of confidence' in the UK economy with real actions. Tianjin Airlines, a subsidiary of China's Hainan Airlines Group, opened a direct flight route connecting Tianjin, Chongqing and London. Sichuan Guodong Construction Group announced its plan to invest £220m in housing projects in Sheffield over the next three years. China Gezhouba Group has also expressed an interest in the UK's water conservancy projects, to the tune of about £1bn. I hope that Britain will continue to be pragmatic and stay open to Chinese businesses … China always takes a strategic and long-term perspective toward its relationship with the UK. I hope the UK will do the same and our two countries will join hands to write a new chapter for China-UK relations". As argued before, there are plausible reasons why this rather optimistic view must not necessarily become reality.

In terms of Brexit-related homework, the British government without a doubt has to prioritise the negotiations with the EU. But as Liddle (2017) noted, it is also important that the UK rethink its future trade relationship with China with the aim to achieve an advantageous trade agreement. The UK government may build its post-Brexit relationship with China on its ties with Hong Kong which includes a double

[32] *The Telegraph*, "Hammond says Brexit Britain must back China's new Silk Road", 14 May 2017, available at http://www.telegraph.co.uk/news/2017/05/14/hammond-says-brexit-britain-must-back-chinas-new-silk-road/.

[33] *The Telegraph*, "China and UK can write new chapter together", 16 September 2016, available at http://www.telegraph.co.uk/news/world/china-watch/politics/china-uk-relationship/.

tax treaty. Furthermore, the UK also has a double taxation agreement (DTA) with mainland China which needs to be updated.

11.3.2 UK–China Trade in the Post-Brexit Era and the Future of the Yiwu–London Train Service

In 2015, the British government at the time announced its ambitious target of elevating China to the second largest trading partner after the USA by 2025.[34] In Chinese government and business circles, there are rather positive expectations about economic relations with the UK post-Brexit, with hopes for free trade, especially in financial services. The invitation for British Prime Minister Theresa May to attend the Belt and Road Forum in May 2017 can be interpreted as a sign that the UK will play an important role in promoting and implementing the BRI after Brexit. May did not eventually attend the forum. Anyhow, it remains unclear what kind of cooperation could be behind optimistic statements such as "there will be a lot of room for collaboration between the two countries".[35] Other voices sound less optimistic. Brown (2017) expects a more complex post-Brexit position because UK is not in the centre of China's interests in Europe as the China-led 16+1 format indicates. This also holds for the 21st Maritime Silk Road.

Summing up, a number of Chinese statements give the impression that the UK is considered to be a true supporter of free trade who has been hampered by the EU during its membership. The British government views the BRI as a welcome "vehicle" to create new trade. The assumption that Brexit will have a trade-creating effect depends essentially on the content of new trade arrangements. Evidence suggests that the British government aims to negotiate free-trade agreements with a host of countries, starting with Australia/New Zealand, USA and

---

[34] *The Guardian*, "Xi Jinping's state visit: China's relationship with Britain in numbers", 20 October 2015, available at https://www.theguardian.com/news/datablog/2015/oct/20/xi-jinping-state-visit-chinas-relationship-with-britain-in-numbers.

[35] CNBC, "Brexit to open new opportunities in China-UK trade", 25 March 2017, available at https://www.cnbc.com/2017/03/25/uk-china-trade-brexit-li-ruogo-boao-forum.html.

China.[36] In addition to the open question of the time horizon (and a UK-proposed transition deal), it is uncertain whether the UK's negotiating position is strong enough to achieve favourable results. Finally, the benefits of the Silk Road Economic Belt depend on factors including the content of a trade arrangement with the EU as the China–UK train route passes through at least four EU member states (Poland, Germany, Belgium, and France). In this respect, an important part of the agreement refers to transit regulations.

With respect to the future total UK–China trade volume, one has to take into account the macroeconomic performance of the UK, the EU and China. The expected slowdown in UK GDP growth post-Brexit might reduce Britain's demand for Chinese products. This, in turn, negatively affects the profitability of the Yiwu-London freight train service. Furthermore, the competitiveness of the train route, compared to sea transport, might decrease. First, reduced demand in Britain could increase the transport costs per container unit. Second, tariffs, administrative costs and the waiting time on the Belarus–EU and EU–UK borders have a negative impact on the competitiveness of the China–UK land route. This might also hold true for eastbound trains. Assuming that the price and income elasticities of the UK's exports to China by freight trains are lower than China's exports to the UK, the land route stands to be negatively rather than positively affected by Brexit. Another negative post-Brexit effect could result from the use of German company DB Cargo for hauling containers on the UK-Duisburg leg. It remains to be seen what concrete results are produced by a 2017 multilateral agreement among the Belarusian Railway, the China Railway Corporation, Deutsche Bahn, Kazakhstan Temir Zholy, Poland's PKP, the Russian Railways and Ulaanbaatar Railway. The agreement, signed in late April 2017, aims to offer uniform end-to-end service standards as well as to reduce transit times, coordinate the development of infrastructure including logistics facilities and introduce technologies

[36]According to research by *The Financial Times*, Britain may ultimately renegotiate more than 750 international treaties agreed by the EU. *Financial Times*, "Doubts grow over transition panacea", 20 October 2017, p. 4.

to support electronic exchange of freight data and, last but not least, to simplify border procedures and reduce customs clearance times. It is about the question whether the Duisburg–London leg will need a special arrangement between the German carrier and the British port authorities in the post-Brexit era. These and other unexpected effects of Brexit on the Yiwu–London freight train route may re-evaluate the competitiveness of the sea route despite the fact that the Yiwu–London train service is around 30 days faster than ocean-going ships on the southwest passage. Finally, and most important, due to the limited transport capacity of trains compared to ships, the former will not gain more than a small fraction of the total transport volume between China and the UK. This in particular applies to eastbound trains as their pre-Brexit frequency is much lower than that of westbound trains. British products exported on the London–Yiwu train—pharmaceuticals, vitamins, and soft drinks—could become less competitive due to administrative barriers when passing through the EU. Furthermore, it is to be expected that companies will relocate production from the UK to EU member states. This will mainly affect machinery and road vehicles which are the UK's top exports to China.[37] Finally, it is worth mentioning that CO_2 emissions from ships on the ocean route are roughly half the emissions generated by the Yiwu-London land transportation service.[38]

There are hardly convincing arguments that the Yiwu–London rail route will have a bright future. In the end, this route might stand symbolically rather than economically for the modern land ties between China and the UK. This conclusion is supported by China's efforts to develop a China–Europe Arctic Sea route. In a recent report by Radio

[37]BMW recently announced a plan to build its electric Mini model in the UK from 2019 to 2023. Parts and components would be produced in Germany and exported to the UK. However, the company acknowledged the risk of tariffs on parts and components imported to the UK and on cars exported to the EU, saying it might eventually be forced to relocate production to Germany unless there was a free-trade agreement. *Fortune*, 25 July 2017, available at http://fortune.com/2017/07/25/bmw-will-make-the-electric-mini-in-the-u-k-despite-brexit-risk/.

[38]Railway technology, available at http://www.railway-technology.com/features/featureeast-wind-a-new-era-of-freight-between-the-uk-and-china-5740643/.

Canada International (RCI) entitled "China's Arctic Road and Belt Gambit",[39] it is argued that China needs to push forward the Arctic route due to the vulnerability of its current trade routes as well as to concerns regarding negative effects on climate change. This northwest sea route from Shanghai to Rotterdam needs approximately 10–12 days longer travel time compared to the China–Europe rail routes, but it is around 15 days shorter than the south-west sea route. It does not make much difference in terms of travel time whether the destination of the Arctic route is Rotterdam or London. The main advantage of this route is that it bypasses the EU. This would make it possible to realise the "declare at home, release at destination" clearance model in its purest form. Statements from China Ocean Shipping Company (COSCO) confirm the company's interest in developing this Arctic Sea route. The COSCO-UK CEO confirmed that "the company has seen a substantial increase in its sea shipping business through the London Gateway—the deep sea port in east London—since the Belt and Road Initiative got underway" (ABP 2017, p. 7), and the COSCO-China CEO added that "COSCO Shipping is optimistic about the future of the NSR [Northern Sea Route] Arctic shipping".[40] The disadvantages of the Arctic Sea route are the dependence on weather conditions in the winter season and therefore higher supply chain risks such as disruptions and delays. In case of a post-Brexit shift of British manufacturing to continental Europe, this argument will lose significance.

Considering the UK's mere 2% of China's total trade, it is difficult to imagine that the Chinese government will put great efforts into negotiations with the UK and the EU to achieve favourable transit conditions for its container shipments via the Yiwu–London train route. This might be different from China's interest in future investment projects in the UK.

[39]RCI, 3 October 2017, available at http://www.rcinet.ca/eye-on-the-arctic/2017/10/03/chinas-arctic-road-and-belt-gambit/.

[40]Quoted in *The Independent Barents Observer*, 10 October 2016, available at https://thebarentsobserver.com/en/arctic-industry-and-energy/2016/10/cosco-sends-five-vessels-through-northern-sea-route.

11.3.3 Post-Brexit BRI Investment in UK

China's preference for investment in the UK has several reasons. It starts with English as the most widespread foreign language in China. Furthermore, a considerable proportion of Chinese graduates who study in the UK return to China with positive impressions. Last but not least, the British history of Hong Kong radiates through manifold channels into present-day mainland China. Anyhow, there are several reasons for a less bright future for Chinese investment in the UK.

First, it is necessary to separate business-oriented FDI from BRI-related infrastructure investment. Regarding the former, new Chinese rules for OFDI by private companies (Table 11.1) will most likely have a negative impact on FDI inflows from China. According to new guidelines on overseas investments, jointly published on 18 August 2017 by the NDRC and three other government institutions (NDRC, MOFCOM, PBOC and MFA 2017), overseas investment in "irrational" acquisitions of assets in industries such as real estate, entertainment, hotels and sports clubs is restricted unless aligned with China's national development, macroeconomic, international cooperation and foreign policies. In recent years, such projects have been the focus of Chinese investment in the UK. But the Brexit decision puts risks on this kind of investment. The decline of the British pound against the RMB devalues "old" Chinese investments. At first sight, a low British pound may attract new Chinese investment. But asset-seeking OFDI is now restricted by the above-mentioned Chinese guidelines. In the past, this kind of OFDI accounted for 40–50% of total Chinese OFDI in the UK (Kratz 2016). Investors seeking high returns will probably look for opportunities in other countries. Chinese companies in manufacturing sectors that plan to establish European operational hubs and production networks will face Brexit-related non-tariff trade barriers (NTBs) or, in the worst case, even tariffs between the UK and the EU. Therefore, the UK will lose some of its attractiveness for Chinese manufacturing companies looking for efficiency gains and/or market access to the EU27. Countries in CEE offer alternative locations. Chinese technology-seeking OFDI may be least affected by Brexit.

In all, the strong decline of Chinese investment in the UK in the first half of 2017 is an early sign of re-adjustment in China's OFDI in the post-Brexit era. On 3 November 2017, the NDRC published a draft revision of the "Administrative Rules for Outbound Investments by Enterprises" for public consultation (Covington & Burling 2017b; Linklaters 2017). Shortly afterwards, on 18 December 2017, the NDRC issued guidelines for outbound private investment that aim to address negative effects such as neglect of quality and safety. According to the document, "private businesses should improve internal rules on decision-making and financial management in terms of overseas investment, and strengthen risk control by using safety measures and contingency plans".[41] The 33.5% year-on-year drop in non-financial OFDI from January to November 2017 can be viewed as an announcement effect of the tighter regulations. The NDRC announced that similar rules for state-owned enterprises were in the pipeline. This would raise the question of how to evaluate the Chinese investment in the London Royal Albert Dock project. China's CITIC Group Corporation as the constructor is a state-owned enterprise, whereas one of the buildings that serve as a promotion centre for Chinese brands will be built by a private Chinese company from Guangdong province. It remains to be seen whether the project comes into conflict with any new guideline.

Contrary to this new restrictive policy, the Chinese government is encouraging projects linked to the BRI. Evidence suggests that the UK does not offer a wide range of infrastructure projects that fit into the initiative. The Hinkley Point C nuclear power plant project as well as the two other planned projects (see below) may face some problems in the post-Brexit era. Brexit means that the UK has to leave the European Atomic Energy Community (EURATOM). As a consequence, the UK needs to introduce new safety inspection regimes. Furthermore, the UK will be excluded from EURATOM nuclear fusion research, nuclear supply chains, equipment and trained staff. The Chinese contractor needs to adjust to new UK standards. The inclusion of France's

[41]Quoted in *China Daily*, 18 December 2017, available at http://www.chinadaily.com.cn/a/201712/18/WS5a37841ea3108bc8c67356b2.html.

EDF may make the situation even more complicated. After the UK passed the "test of mutual trust" (Chinese ambassador to the UK) with the approval of Chinese investment in the Hinkley nuclear power plant in September 2016, China feels encouraged to build further nuclear power plants in the UK and even worldwide. The Shenzhen-based General Nuclear Corporation is working to get approval from the British government for its third-generation HPR1000 nuclear reactor. According to *China Daily*,[42] the company is confident of getting through the ongoing government assessment. After the approval of the Hinkley Point C project, the Bradwell B plant in Essex would be another Chinese investment in UK energy infrastructure. Work to build the plant was originally scheduled to start in 2022 or 2023, but the timetable has not been confirmed so far. In addition to passing the Generic Design Assessment, the Chinese company would need planning consent and a subsidy contract from the British government. Furthermore, there is an understanding that China will be involved in the development of a third nuclear power plant (Sizewell C) provided that Bradwell B will be approved by the UK government. The Chinese-led comeback of nuclear energy in Britain may revive the discussion on the UK's national security.

Overall, while the future of trade and investment post-Brexit is uncertain, the future of London as a financial centre is expected to be rather positive, not least due to the BRI. "London will remain an important financial centre under any plausible circumstances. It survived the 1930s and two world wars. It will survive Brexit. Yet, within the EU, it was emerging as the undisputed financial capital of Europe, as well as one of the world's two most important financial centres. After Brexit, it is likely to become an offshore centre, relatively more vulnerable to policy decisions, especially regulatory decisions, made elsewhere".[43] As mentioned before, London plays an important role in

[42]*China Daily*, 26 August 2017, available at http://www.chinadaily.com.cn/business/2017-08/26/content_31137906.htm.

[43]M. Wolf quoted in *The Financial Times*, 9 July 2017, available at https://www.ft.com/content/63e937b0-44fd-11e6-9b66-0712b3873ae1.

financing BRI projects in all participating countries as well as in internationalising the renminbi. But there are also warning voices. Some commentators point out that London's attractiveness in terms of finance depends on the arrangement with the EU. If British negotiators are not successful in negotiating a special deal for financial services, the City of London will hardly keep its leading position as a hub for Chinese banks and investors. It is to be expected that their focus will be on doing business within the single market.[44] Chinese banks seem to be delaying their decisions to invest in London until clarity on the regulatory frameworks in the UK and the EU has been achieved. In the short run, the possibility to shift BRI financing from London to other financial centres is limited. If major global banks move their capital market operations out of London, China needs to look for an alternative, at least for its BRI financing.

11.4 Conclusions

At the time of President Xi Jinping's state visit to the UK in 2015, the Chinese foreign minister stated that "China hopes to see a prosperous Europe and a united EU, and hopes Britain, as an important member of the EU, can play an even more positive and constructive role in promoting the deepening development of China-EU ties".[45] There is sufficient evidence that China will adjust its BRI strategy towards the UK for several reasons. *First*, the UK will lose its preferred location for entering markets in the EU. This in particular applies to investment in manufacturing, but also to trade if British companies or companies with headquarters in the EU27 relocate production to the continent and China is part of the transcontinental value chain. Further unchanged market access to the EU27 is vital for China's economy, especially in

---

[44] *Business Insider*, 6 September 2017, available at http://www.businessinsider.de/uk-china-trade-deal-after-brexit-2017-8?r=UK&IR=T.

[45] Quoted by Yu Jie, LSE blog, available at http://blogs.lse.ac.uk/brexit/2015/11/23/only-continued-membership-of-the-uk-in-the-european-project-can-ensure-a-long-lasting-strategic-partnership-with-china/.

view of slowing growth rates. In the post-Brexit era, China's BRI strategy towards Europe may lead to more and faster infrastructure investment in CEE, which may motivate Chinese companies to reallocate production to this region. Furthermore, BRI-financed upgrading of freight train destinations may contribute to an increased trade volume on land routes. China–UK trade may be redirected from land transportation to the Arctic Sea passage. *Second*, BRI investment in the UK's energy sector is not yet cut and dried. Anyhow, China needs to demonstrate the functioning of its new nuclear power plant technology in order to win contracts in other countries. This could lead to political bargaining processes in the context of a China–UK trade and investment agreement. With respect to the London Royal Albert Dock project, legal complications could occur depending on the rules and regulations of a code of conduct for OFDI by SOEs to be published in the near future. *Third*, it is to be expected that London as the top-tier financial centre will lose some of its attractiveness. In the short run, China must remain confident about the capability of the City of London to significantly contribute to the BRI's financial needs as well as to push forward the internationalisation of the renminbi. Large Chinese banks seem to be hesitant to rapidly expand their presence in London. It is not surprising that the British government relies on the strength of the UK in financial services and will therefore play a key role in the future of the BRI (Fairhead 2018).

It is to be assumed that any adjustment in China's BRI activities in the UK will also lead to changes in its activities in the EU27. This would be most obvious in trade and transportation infrastructure. Countries in CEE could be the main beneficiaries. On the one hand, even though the BRI is a long-term project, China needs to present successes in its initial stage. This is a strong indication that China will adjust its post-Brexit BRI strategy towards the UK where it considers that vital to its interests, but keep the status quo in all other cases. On the other hand, the UK will need new partners after Brexit and China's BRI might offer a promising anchor. There is evidence that the British government did not convincingly succeed in communicating its Brexit vision, which includes an upgraded relationship with China without being perceived as a "Chinese vassal in Europe", as Kerry Brown (2018) put it.

Finally, it should be borne in mind that a comprehensive report on China's BRI activities in and with the UK requires deeper research in all fields of cooperation. Limits are set, first, by the hardly available data (which is partly due to the short period of observation between the end of 2015 and the end of 2017, with the Brexit vote in mid-2016); second, by the high dependence on (sometimes contradictory) statements/reports in non-scientific sources; and, third, by the mixing between purely commercial projects and BRI-related projects. Last but not least, the uncertainty about the Brexit conditions makes it uncertain how China will react.

References

ABP. (2017, October). Construction Begins on £1.7BN Royal Albert Dock Project. *Quarterly Review, 1,* 2–4.

Blanchard, J.-M., & Flint, C. (2017). The Geopolitics of China's Maritime Silk Road Initiative. *Geopolitics, 22*(2), 223–245.

Brown, K. (2017, May 12). How Brexit Britain Can Gain from China's Belt and Road. *South China Morning Post.* Available at http://www.scmp.com/week-asia/opinion/article/2094166/what-brexit-britain-has-gain-chinas-belt-and-road.

Brown, K. (2018, January 30). Britain's China Challenge. *The Diplomat.* Available at https://thediplomat.com/2018/01/britains-china-challenge/.

Casaburi, I. (2016). Chinese Investment Trends in Europe: 2016–17 Report. *ESADE China-Europe Club.* Barcelona.

China-British Business Council. (2016). *China-Britain Belt and Road—Case Studies Report—Global Partnerships in Action.* London.

Covington & Burling. (2017a, September 8). *China's State Council Publishes New Guidance on Regulation of Outbound Investment.* Beijing.

Covington & Burling. (2017b, November 24). *NDRC Releases Draft Enterprise Outbound Investment Regulations for Public Comment.* Beijing.

Dhinga, S., & Sampson, T. (2016). *Life After Brexit: What Are the UK's Options Outside the European Union?* London: Centre for Economic Performance.

Emerson, M. (2016). *Which Model for Brexit?* CEPS Special Report. Brussels: Centre for European Policy Studies.

European Parliamentary Research Service. (2016, July). One Belt, One Road (OBOR): China's Regional Integration Initiative. *Briefing.* European Union, Brussels.

EY. (2017, May). *EY's Attractiveness Survey UK—Time to Act*. London.

Fairhead, R. (2018, May 10). Why China's Belt and Road Offers the UK Huge Opportunities. *CapX*. Available at https://capx.co/why-chinas-belt-and-road-offers-the-uk-huge-opportunities/.

Fues, T. (2017). *Belt and Road Initiative: Emergence of a Parallel International Order?* German Development Institute. Available at https://www.die-gdi.de/fileadmin/user_upload/pdfs/Ausbildung/20170406_Fues_Belt_and_Road_Initiative_April_2017.pdf.

Garcia-Herrero, A. (2017, May 23). China Can't Finance 'Belt and Road' Alone. In Marsh & McLennan (Ed.), *Navigating the New Silk Road—Expert Perspectives on China's Belt and Road Initiative* (pp. 13–16). Brink Perspective.

Global Counsel. (2015). *Brexit: The Impact on the UK and the EU*. London. Available at https://www.global-counsel.co.uk/sites/default/files/special-reports/downloads/Global%20Counsel_Impact_of_Brexit.pdf.

Hanemann, T., & Huotari, M. (2018, May). *EU-China FDI: Working Towards Reciprocity in Investment Relations* (Papers on China No. 3). Berlin: Mercator Institute for Chinese Studies.

Kaplan, Z. (2016). *EATL: The Trade Prospects for the EU and China*. UNECE. Available at https://www.unece.org/fileadmin/DAM/trans/doc/2016/wp5-eatl/WP5_GE2_2nd_infromal_session_Ms_Kaplan_1.pdf.

Kofner, Y. (2018, July 15). The Chinese Belt and Road Initiative (BRI). *Eurasian Studies*. Available at http://greater-europe.org/archives/5596.

Kratz, A. (2016). Brexit's Impact on Chinese Investments in the UK. *China Policy Institute: Analysis*. Available at https://cpianalysis.org/2016/08/19/brexits-impact-on-chinese-investments-in-the-uk/.

Le Corre, P. (2017, May 23). Europe's Mixed Views on China's One Belt, One Road Initiative. *Brookings*. Available at https://www.brookings.edu/blog/order-from-chaos/2017/05/23/europes-mixed-views-on-chinas-one-beltone-road-initiative/.

Liddle, K. (2017, March 3). China-UK Trade: The Effects of Brexit. *China Briefing*. Available at http://www.china-briefing.com/news/2017/03/03/china-uk-trade-the-effects-of-brexit.html.

Linklaters. (2017, November 9). *China's New Approach to Regulation of Outbound Investment*. Shanghai. Available at https://www.linklaters.com/de-de/insights/publications/asia-news/china-update/2017/chinas-new-approach-to-regulation-of-outbound-investment.

Liu, D., Gao, H., Xu, Q., Li, Y., & Song, S. (2017a). *The 'Belt and Road' Initiative and the London Market—The Next Steps in Renminbi*

Internationalization. Part 1: The View from Beijing. London: Chatham House and The Royal Institute of International Affairs.
Liu, D., Gao, H., Oxenford, M., Xu, Q., Song, S., Subacchi, P., & Li, Y. (2017b). *The 'Belt and Road' Initiative and the London Market—The Next Steps in Renminbi Internationalization. Part 3: Framework for Policy Discussion.* London: Chatham House and The Royal Institute of International Affairs.
Lobyrev, V., Tikhomirov, A., Tsukarev, T., & Vinokurov, E. (2018). *Belt and Road Transport Corridors: Barriers and Investments* (MPRA Papers, No. 86705). Available at https://mpra.ub.uni-muenchen.de/867056/.
Mercator Institute for China Studies. (2018, June 7). *Mapping the Belt and Road Initiative: This Is Where We Stand.* Available at https://www.merics.org/de/bri-tracker/mapping-the-belt-and-road-initiative.
NDRC. (2015). *Vision and Actions on Jointly Building Silk Road Economic Belt and 21st-Century Maritime Silk Road.* National Development and Reform Commission. Available at http://en.ndrc.gov.cn/newsrelease/201503/t20150330_669367.html.
NDRC, Ministry of Commerce (MOFCOM), People's Bank of China (PBOC) and Ministry of Foreign Affairs (MFA). (2017, August 18). *Notice on Further Guiding and Regulating the Directions of Outbound Investment.* Beijing. Available at https://www.cov.com/-/media/files/corporate/publications/2017/09/chinas_state_council_publishes_new_guidance_on_regulation_of_outbound_investment.pdf.
OBOReurope. (2018, March 8). *European Companies Show Willingness to Participate in the BRI.* Available at http://www.oboreurope.com/en/european-companies-bri/.
OLGBIR. (2017). *Building the Belt and Road: Concept, Practice and China's Contribution.* Beijing: Foreign Languages Press.
PwC. (2016). China's New Silk Road: The Long and Winding Road. *PwC's Growth Market Centre.* Available at https://www.pwc.com/gx/en/growth-markets-center/assets/pdf/china-new-silk-route.pdf.
Rodriguez, J. A. (2018). *The Belt and Road Initiative: A Geopolitical Approach* (BA thesis in Chinese Studies). University of Iceland. Available at https://skemman.is/bitstream/1946/29434/1/Jos%C3%A9%20Alam%C3%A1%20Rodr%C3%ADguez.pdf.
Schubert, J. (2017). *New Eurasian Age: China's Silk Road and the EAEU in SCO Space.* Available at http://russianeconomicreform.ru/wp-content/uploads/2017/08/New-Eurasian-Age-with-Chinese-Silk-Road-and-EAEU-in-SCO-Space-Update.pdf.

SCMP. (2017, August 18). *Overseas Real Estate and Soccer Clubs in Crosshairs as China Ramps Up Investment Clampdown*. Available at http://www.scmp.com/news/china/economy/article/2107392/overseas-hotels-and-soccer-clubs-crosshairs-china-ramps.

State Council, People's Republic of China. (2017, August 18). *State Council Issues Guideline on Overseas Investment*. Available at http://english.gov.cn/policies/latest_releases/2017/08/18/content_281475798846134.htm.

Steer Davis Gleave. (2018). *Research for TRAN Committee: The New Silk Route—Opportunities and Challenges for EU Transport*. Brussels: European Parliament, Policy Department for Structural and Cohesion Policies.

Subacchi, P., & Oxenford, M. (2017). *The 'Belt and Road' Initiative and the London Market—The Next Steps in Renminbi Internationalization. Part. 2: The View from London*. London: Chatham House and The Royal Institute of International Affairs.

Summers, T. (2016). The United Kingdom: A Platform for Commercial Cooperation. In F. P. van der Putten, J. Seaman, M. Huotari, A. Egman, & M. Otero-Iglesias (Eds.), *Europe and China's New Silk Roads: A Report by the European Think-Tank Network on China* (pp. 63–68). Den Haag: Clingendael Institute.

Suokas, J. (2018, March 8). China-Europe Rail Cargo Grows Dramatically But Challenges Remain. *gbtimes*. Available at https://gbtimes.com/china-europe-rail-cargo-grows-dramatically-but-challenges-remain.

SWIFT. (2018, January). RMB Internationalisation: Where We Are and What We Can Expect in 2018. *RMB Tracker*. La Hulpe, Belgium.

UNCTAD. (2017). *World Investment Report 2017—Investment and the Digital Economy*. Geneva: United Nations Publications.

Ward, M. (2018). *Statistics on UK Trade with China* (Briefing Paper No. 7379). House of Commons Library.

Wye, R. (2016). China Would Not Abandon UK After Brexit. *Chatham House*. Available at https://www.chathamhouse.org/expert/comment/china-would-not-abandon-uk-after-brexit.

Zhang, X. (2016). Chinese Capitalism and the New Silk Roads. *Aspen Review, 4*. Available at https://www.aspenreview.com/article/2017/chinese-capitalism-and-the-new-silk-roads/.

Zhao, H. (2016). China's One Belt One Road: An Overview of the Debate. *Trends in Southeast Asia, 6*(16), ISEAS-Yusuf Ishak Institute, Singapore.

12

The Impact of Brexit on Foreign Direct Investment and Trade Relations Between the UK and China

Marta Mackiewicz and Agnieszka McCaleb

12.1 Introduction

The main aim of this chapter is to identify the possible impact of Brexit on foreign direct investment and trade relations between the UK and China. The main research questions are: (i) How will Brexit affect Chinese foreign direct investment in the UK? (ii) Can the UK benefit from shaping new trade relations with China?

A study of a potential Brexit effect on selected aspects of the global economy must be reduced to some general considerations due to a wide range of possible Brexit scenarios and possible developments that are difficult to predict even for the parties directly involved. It is even more difficult to make predictions regarding the UK's future relationship

M. Mackiewicz · A. McCaleb (✉)
Collegium of World Economy,
SGH Warsaw School of Economics, Warsaw, Poland
e-mail: amccal@sgh.waw.pl

M. Mackiewicz
e-mail: mmackie@sgh.waw.pl

A. M. Kowalski (ed.), *Brexit and the Consequences for International Competitiveness*, https://doi.org/10.1007/978-3-030-03245-6_12

with non-EU countries. The intended process of departure from the European Union is planned in two steps. The first stage of the "divorce" is related to the new EU-UK relationship. This will in due course affect Britain's relationship with non-EU countries, including in trade and investment.

The research methods used for answering the research questions are a literature review and a quantitative analysis of statistical data on trade and investment. Research reports and news articles on Brexit naturally focus on the UK. This imposes a skewed way of thinking about the future of British-Chinese trade relations: what are the strengths and weaknesses of the British economy, and how will the country's future relationship with the EU affect these strengths and weaknesses. A bit neglected—but potentially just as important—factor is the shape of future relations between the EU and China and its impact on UK-China relations. In fact, we are dealing with a triangle relationship, with each side affecting the remaining two. In the simplest terms, the EU will become a competitor for the UK in a different way than before. Moreover, the rest of the EU will be a more important partner for China, so the British-Chinese relationship may be somehow secondary to EU-China relations. Given the significant UK contribution to the current EU policy towards the Middle Kingdom, the divergence between what the EU and the UK are striving for is likely to deepen, especially in terms of a free-trade agreement, market economy status and an investment agreement.[1] Another factor limiting the freedom of shaping the future relationship will be relations with the USA.[2]

This chapter is organised as follows. The first part discusses the implications of Brexit for Chinese foreign direct investment in the UK. The next part focuses on the exchange of goods and trade in services. It analyses Eurostat data on exports and imports and offers an

[1]Summers, Tim, *Brexit: Implications for EU-China Relations*, Royal Institute of International Affairs, Chatham House, 2017.

[2]Summers, Tim, *Brexit and the UK's China Challenge*, https://www.chathamhouse.org/expert/comment/, 2016.

analytical concept to compare the structure of trade with and without trade restrictions. On the basis of this analysis, possible scenarios are developed. The last part concludes.

12.2 The Impact of Brexit on Chinese OFDI in Europe

Since the beginning of the 2000s, China's outward foreign direct investment (OFDI) has been growing rapidly, making the country the world's third-largest source of FDI (UNCTAD/WIR 2017). Europe attracted Chinese FDI motivated mainly by strategic asset seeking, i.e. brands, technology, know-how, R&D infrastructure and distribution channels through which Chinese firms improved their competitiveness in the global and domestic markets. Another benefit was access to the EU market, which is the number one recipient of Chinese exports and the largest market in the world. In 2016, the highest growth in the value of Chinese investment transactions in the EU28—compared to the annual average for 2013–2015 period—was recorded for industrial machinery and equipment, ICT, utilities, transport and infrastructure.

Within the EU, the UK has been the largest recipient of Chinese outward FDI, followed by Germany, while also serving as an investment base for Chinese investment in other member states (Hanemann and Huotari 2017).

Chinese investors in the UK are particularly interested in real estate (44% of the total value of deals between 2012 and the first half of 2016), energy (7%), finance (8%), health care (5%), agriculture, technology and entertainment industries (American Enterprise Institute and the Heritage Foundation 2017; Kynge 2017). High Chinese investment in British real estate is a result of Chinese businesses' motivation for safe placement of their assets. In the 2016–2017 fiscal year, the number of FDI projects carried out by Chinese investors as well as the number of newly created jobs increased compared with the previous year (UK Trade and Investment Inward Investment Reports 2017). It is important to note that the number of Chinese FDI projects in the UK in the period 2016–2017 was higher than in the preceding years despite restrictions on outward FDI introduced by the Chinese government at the end of 2016 (which was a record year for Chinese OFDI, seeing it grow by 30% in year-on-year terms to USD 188.8 billion). The restrictions were motivated by falling value of country's currency (Renminbi) and decreasing foreign exchange reserves (Davies 2016; Ernst & Young 2017). In 2016, the UK ranked fourth globally in terms of Chinese M&As (Liu 2017). In terms of the number of Chinese M&A deals in the EU, the UK has been leading the way, implying a strong interest in strategic assets and much less interest in market expansion through FDI (Clegg and Voss 2012). This is confirmed by OECD findings that China supplies the UK mainly by trade as opposed to the USA, France, the Netherlands and Japan, which do so through trade and sales by foreign affiliates (OECD 2017). Chinese investors are not only owners of renowned UK brands such as House of Fraser, MG Rover, Pizza Express and Weetabix (Voss 2017), but they also invest in R&D centres benefitting from specialised clusters, such as Sinovet, which in 2015 established a new animal health R&D facility near Edinburgh (UK Trade & Investment 2015).

Based on the above, it can be inferred that Chinese investment in the UK will not be greatly affected by Brexit as their motivations are mostly related with securing assets (real estate investments) and strategic-assets seeking. In fact, British firms became cheaper for Chinese businesses as the British pound lost value against the Chinese Renminbi: falling from 9.43 on the day of referendum to 8.97 in May 2017 and 8.76 in November 2017. Meanwhile, the euro gained more than 5% over that time, as a result of which euro-denominated assets became more expensive for Chinese investors (Voss 2017). The number of newly announced M&A deals dropped by 20% in the first six months of 2017 compared with the same period in 2016 (Hanemann and Huotari 2017). The slowdown in Chinese corporate FDI might be explained by a wait for further devaluation and better deals. Chinese motivations after Brexit will remain the same, notably a search for prominent brands and cutting-edge technologies. Chinese market-seeking firms that target the whole EU market will most probably lose interest in the UK.

12.3 The Impact of Brexit on Trade Relations Between the EU, the UK and China

Worthy of mention are the British government's assumptions about the rules shaping future trade policy, in particular the pursuit of high consumer, employee and environmental protection standards in trade agreements.[3] Too many requirements in these areas might limit the freedom to form relationships with China. Finally, it is worth mentioning the Belt and Road Initiative (BRI), an important aspect of Chinese policy in which Britain plays a rather marginal role. This limited involvement may hamper negotiations on trade relations.[4]

All these issues will be crucial for UK-China trade in the long term. Meanwhile, in the short and medium term, two macroeconomic developments will affect trade. The weakening of the pound has already made British exports cheaper and imports from China have become more expensive. This effect has been changing trade flows ever since the Brexit referendum. Second, as rightly noted in Rothman 2016, a broader anxiety over Brexit may induce a material economic downturn in both the EU and Britain.[5] That would affect both the size and composition of trade with China.

For the UK, China was in 2015 the fourth-biggest recipient and second-largest seller of goods, constituting 5.9 and 9.9% of total British trade flows respectively. With respect to services, China was in third and seventh place respectively (1.4 and 1.0% of the total). It is worth noting that Hong Kong is also a major partner in the trade of services.[6] In goods trade, China is one of the main partners, but its role in the exchange of services is less significant.

From China's point of view, the UK is a less significant partner. In the trade of goods, the UK is the ninth largest export market (2.6% in

---

[3]Department for International Trade, Preparing for Our Future UK Trade Policy, p. 29, October 2017.

[4]Brown, Kerry, *How Brexit Britain Can Gain from China's Belt and Road*, http://www.scmp.com/week-asia/opinion/article/2094166/what-brexit-britain-has-gain-chinas-belt-and-road, 2017.

[5]Rothman, Andy, *Brexit Impact on China*, Advisor Perspectives, 2016.

[6]Based on comtrade.un.org.

2016) and the 20th largest import market (1.2%) for China.[7] This disparity may be an important factor shaping the future trade relationship between the countries.

Future commercial relationships within the triangle will be shaped by the relative role of the UK and other EU member states in trade with China. In this context, it can be noted that the UK is China's second-largest export market in the EU. The share of the UK in the trade of goods was 16.7% in 2015, lower than Germany's and equal to that of the Netherlands. It can be concluded that the position of the UK is important in this dimension. Meanwhile, Britain's role in exports from the EU to China is less significant. The UK was the third-largest exporter in 2015, with a share of 9.2% vs. Germany's more than 42%.[8] This considerable disproportion has been cited as a confirmation that Britain has untapped potential that could be released by greater freedom in shaping trade regulations after Brexit. In terms of the total EU deficit in goods trade with China, the UK is in second place, with a share of 27% vs. Germany's 34%.

The dynamics of the trade relationship (in the exchange of goods) between China and the UK is ambiguous. During the 2010–2015 period, UK imports from China increased by about 53.7%, vastly outperforming the dynamics of the remaining EU countries (8.2%).[9] The same is true of exports where the respective figures were 67.5 and 19.6%. Britain compares unfavourably with other EU countries in terms of the trade deficit. In the case of the UK, the trade deficit increased by close to 50% from 2010 to 2015, while in other countries it fell by 7%.

Other EU countries can be a point of reference for analysing British trade. However, for the full picture, it is worth taking a look at Switzerland, a country whose status is similar to that Britain may obtain after leaving the EU. Switzerland maintains close relations with the EU as a member of EFTA, and at the same time, it is free to negotiate trade

[7]Based on info.hktdc.com.

[8]Own calculations based on stats.gov.cn.

[9]Own calculations based on www.stats.gov.cn.

agreements with third countries separate from the EU. Switzerland has signed 38 such agreements, including one with China.[10] Such a scenario is viewed as potentially beneficial for British exports.[11] For China, the ability to negotiate agreements with European countries independently of EU policy may be a bargaining point in negotiations with the EU, which could potentially allow the UK to secure favourable conditions.[12] Data on trade between Switzerland and China appears to validate such a scenario. Switzerland has a surplus in goods trade, with exports growing by more than 140% from 2010 to 2015 and slow growth of imports. However, the freedom to shape the trade relationship is not the only factor that plays a role. Another important factor is a growing tendency among rich Chinese citizens to buy luxury goods.[13]

The freedom to shape future business relationships will influence UK-China trade insofar Britain's potential is limited by the current EU-China agreement. The extent of limitations resulting from the current regulations and potential changes that may be prompted by Brexit can be analysed in many ways. Below an attempt at a quantitative approach to trade is presented. It is based on an analysis of the current structure of British exports.

The approach applied in the analysis is based on the following reasoning. The starting point is two countries trading under completely unrestricted movement of goods. The structure of commodity exports of one of them to the other is a result of a number of factors, such as the characteristics of the two economies, mutual comparative advantages and geography. Then trade policies are changed and trade is no longer completely free. The likely result of this change will be an adjustment of trade flows. This will probably happen even if the conditions

[10]What consequences would a post-Brexit China-UK trade deal have for the EU?, p. 3, Policy Contribution Issue 18, 2016.

[11]Winders, Sam, *Would a Post-Brexit UK Be Better Able to Sign Free-Trade Agreements with the Rest of the World?* The Bruges Group, 2016.

[12]Summers, Tim, *Brexit: Implications for EU-China Relations*, The Royal Institute of International Affairs, Chatham House, 2017.

[13]For example, the value of watch exports alone (expressed in CHF) increased by about 22% in 2010–2015, or nearly USD 250 million, according to the Federation of the Swiss Watch Industry (FH).

for all items (e.g. uniform customs tariffs) undergo the same changes because the same percentage change in prices does not necessarily cause the same changes in demand in different groups of products. As a result, this changes the structure of the flows. The discrepancy between the initial structure and that following the introduction of restrictions on trade reflects the strength with which the new rules distort trade.

This analytical concept was applied using Eurostat data on British exports in 2016, broken down according to the HS2 nomenclature (around 100 product groups). The structure of trade with China conforms to the revised structure described in the example above. Data on the structure of trade without barriers is obviously unavailable. So an approximation was used for the purpose of this analysis. To simplify the analysis, it can be assumed that the flow of goods within the EU is free, so the structure of Britain's trade with the EU can be used as a point of reference (reference structure)—with certain limitations, as discussed later. The degree of divergence of both structures was determined as follows. For individual commodity groups, the difference in participation in the structure of exports to China and EU countries was determined. The average absolute differences, expressed in percentage points, show the discrepancy between the structures and, by extension, the level of trade distortion:

$$ix = \sum_{i=1}^{n} |s_i - r_i|/n$$

where *ix* denotes the divergence measure index; *n*—number of classes of goods; r_i—share of the *i*-th class of goods in the initial (reference) structure; and s_i—share of the *i*-th class of goods in restricted trade conditions.

In order to determine whether the obtained effect is large or small, a scale is needed. It should also be noted that trade flows, and thus also their structure, may be a result of factors other than the free movement of goods and its restrictions. As a point of reference, analogous indicators for each of the remaining EU countries were used. Any differences between them are due only to the diversity of economies and not the regulation limiting the freedom of trade. For the calculation of these indicators, a modified reference structure was used, determined based

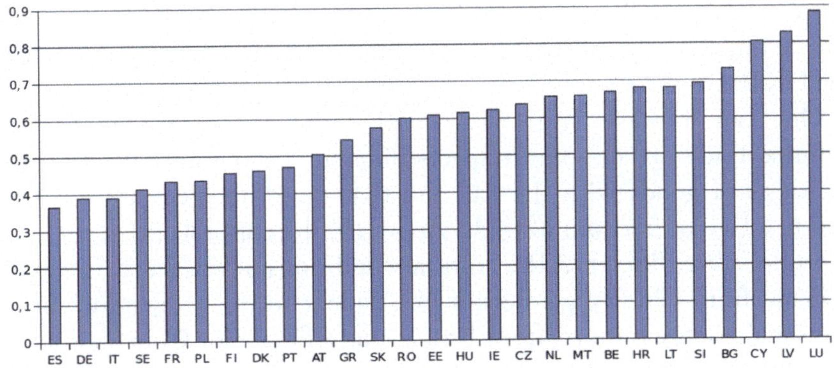

Fig. 12.1 Indicators of structural discrepancy for UK exports to EU countries (*Source* Own calculations based on Eurostat data)

on British exports to EU countries other than the one for which the indicator was calculated. This is due to the fact that large importers have a significant contribution to trade with the EU as a whole and that the value of the indicator would be underestimated.[14] It can be expected that the values obtained by the EU countries will usually be lower than that for China, which results not only from the different nature of the economy but also restrictions on the free movement of goods. The figure below presents the discussed indicators. They range from 36 to 87%, and the larger countries frequently have lower indicators. It can be assumed that this is a consequence of the higher diversity of large economies, more resembling the sum of all EU economies bar the UK. As a result, the weighted average value of the indicator (with the size of British exports to a given country as a weight) is 0.55 p.p., and the arithmetic mean is 0.59 p.p. (Fig. 12.1).

In other words, a large and relatively diverse economy should, under free trade, have an indicator of discrepancy for British exports ranging from 0.37 p.p. to around 0.50 p.p. A higher value may indicate that the existing restrictions on the free movement of trade distort the structure of British exports and suggest the potential for a correction in the case

[14]In other words, for example, in Germany we would compare the structure of German trade with one in which German trade constitutes a significant part.

of free trade post-Brexit. The indicator calculated for China is 0.63 p.p. and so suggests the possibility to improve the matching of trade after the UK leaves the EU.

One can, however, rightly argue that the factors driving British-Chinese trade are in some respects different from those affecting the exchange of goods with the EU and that the differences are not limited to a lack of freedom of trade. Above all, the cultural and geographic conditions are different. The cultural conditions affect the structure of demand from China regardless of free trade, and the geographic conditions make some imports from Britain unprofitable for China, although they are cost-effective for European partners. For these reasons, it is worth analysing the indicators of discrepancy for UK exports to China compared to other non-EU countries (Fig. 12.2).

The indicators range from 0.48 p.p. for Canada to 1.52 p.p. for Switzerland. It is worth noting that not only Canada, but also Australia and the USA have relatively low indicators, below that of China. This can be attributed to cultural and linguistic proximity, which partly compensates for the distortion of trade resulting from geographical factors or restrictions on the free movement of goods. India, a country with lower income, but huge potential and a population comparable

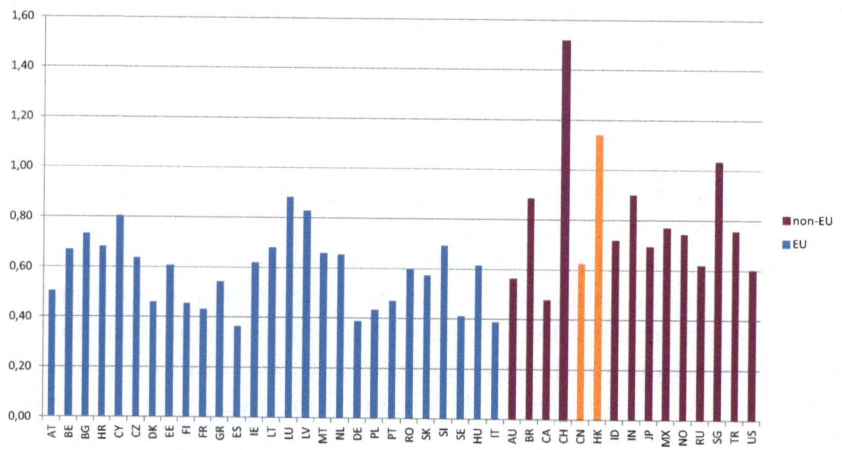

Fig. 12.2 Indicators of structural discrepancy for UK exports to EU and selected non-EU countries (*Source* Own calculations based on Eurostat data)

to that of China, has an indicator of 0.72 p.p. Indonesia, Japan and Brazil also have relatively high indicators. Against the background of all these countries, China's indicator is relatively low, pointing to a limited impact of restrictions on freedom of trade and less potential for a revolution in commercial trade flows because of Brexit. So the benefits of a free-trade agreement with China will at best be modest.[15]

Hong Kong is a different case. Its indicator of 1.14 p.p. may come as a surprise given that just two decades ago the economy was controlled by Britain. However, in this case, the key issue is that Hong Kong is a small economy with a relatively high income per capita. Singapore is another country with similar characteristics (a small Asian city-state with high income per capita) and an indicator of 1.03 p.p.

While this chapter has focused on trade in goods, it is often emphasised that the strength of the British economy lies in services and that the trade of services holds the greatest potential for development after Brexit. The UK is a powerhouse in services such as finance and education, and China is increasingly moving from manufacturing to services. When negotiating free-trade agreements, the European Union must balance various conflicting interests among EU member states, which impairs the free movement of services. However, the argument about the complementarity of China (exports of labour-intensive goods) and the UK (exports of high-tech goods and services) is only seemingly rational. It is worth asking whether this argument is based on a realistic assessment of comparative advantages as China evolves from a low-cost economy to a provider of advanced technology. Such a transformation may help boost British exports in the short term, but in the long run it may prove to be a threat.

The development of services, including financial services, will further deepen the divide between London, which values the country's EU membership, and the rest of Britain.[16] The development of educational

[15]What consequences would a post-Brexit China-UK trade deal have for the EU?, p. 8, Policy Contribution Issue 18, 2016.

[16]Dreyer, Jacob, *Could China Be the New Best Friend for a Post-Brexit Britain?* https://www.newstatesman.com/politics/economy/2017/07/could-china-be-new-best-friend-post-brexit-britain, 2017.

services may be in conflict with the expectations of Brexit supporters and calls to reduce immigration. In context of British service exports, it seems puzzling that there has been little mention of the possibility of the UK wrestling a chunk of the Chinese market away from other partners. Rather, it seems that new demand is expected to appear in China matching what the UK has to offer in various sectors.

12.4 Conclusions

The UK's decision to leave the European Union means that the two partners will have to redefine their political and economic relations. The actual form of divorce between the EU and the UK, either a "soft" or "hard" Brexit, will have significant repercussions for their relations with other major global partners such as China. Despite an initial shock caused by Brexit, it seems that it will have little effect on economic relations between the UK and China. Chinese investors in the UK are attracted by safe investments in real estate and strategic assets, which became cheaper after the British pound took a hit. Meanwhile, other EU locations offering such strategic assets became less appealing as the euro appreciated.

It is often emphasised that British-Chinese trade will benefit from the UK's ability to reshape its international relations, but the common view is that any resulting changes will not necessarily be beneficial in all areas. While the potential for deepening and expanding trade is undoubtedly in evidence, there are many doubts as to whether it is sufficient and whether it can be exploited in the right way. The definitive shape of relations between the EU and a post-Brexit Britain and between the EU and China will be crucial. The quantitative analysis indicates that, although there is potential for the development of UK-China trade, it is probably not as large as is often claimed. The argument for expanding the trade of services is based on factors including the complementarity of what the UK and Chinese service sectors have to offer. But it is uncertain how long this complementarity can last.

It is frequently argued that a post-Brexit UK will be more nimble and free-trade-oriented, which will result in better trade relations. However, it is worth remembering that, for the quality of trade relations, it is equally important to what extent the other side is free-trade-oriented at the same time. Despite frequent declarations (including by Chinese President Xi Jinping at Davos), China's dedication to free trade is unbalanced. It is skewed towards unhindered trade in China-produced goods. So changes in UK-China trade resulting from Brexit may prove to be less significant than theoretically possible.

References

American Enterprise Institute and the Heritage Foundation. (2017). *China Global Investment Tracker*. http://www.aei.org/china-global-investment-tracker/. Accessed November 10, 2017.

Brown, K. (2017). *How Brexit Britain Can Gain from China's Belt and Road*. http://www.scmp.com/week-asia/opinion/article/2094166/what-brexit-britain-has-gain-chinas-belt-and-road.

Clegg, J., & Voss, H. (2012). *Chinese Overseas Direct Investment in the European Union*. Europe China Research and Advice Network. http://www.chathamhouse.org/sites/default/files/public/Research/Asia/0912ecran_clegg-voss.pdf. Accessed November 11, 2017.

Davies, R. (2016). *China to Rein in Outward Investment as Domestic Growth Stalls*. https://www.theguardian.com/business/2016/dec/26/china-to-rein-in-foreign-investment-as-domestic-growth-stalls. Accessed November 11, 2017.

Department for International Trade. (2017). *Preparing for Our Future UK Trade Policy*.

Dreyer, J. (2017). *Could China Be the New Best Friend for a Post-Brexit Britain?* https://www.newstatesman.com/politics/economy/2017/07/could-china-be-new-best-friend-post-brexit-britain.

Ernst & Young. (2017). *EY: China's Outbound Investment in 2017 Is Expected to Steadily Slow Down; Chinese Enterprises Should Concentrate on the Quality of Their Investments to Achieve Stable Development*. http://www.ey.com/cn/en/newsroom/news-releases/news-2017-ey-china-outbound-investment-is-expected-to-steadily-slow-down. Accessed November 11, 2017.

Hanemann, T., & Huotari, M. (2017). *Record Flows and Growing Imbalances, Chinese Investment in Europe in 2016*. Mercator Institute for China Studies.

Kynge, J. (2017). *Strong Growth for Top Chinese Investments in UK.* https://www.ft.com/content/262d1258-0725-11e7-97d1-5e720a26771b. Accessed November 11, 2017.

Liu, C. (2017). *Chinese Companies Will Continue to Look for UK Investments.* http://www.chinadaily.com.cn/business/2017-06/30/content_29942193.htm. Accessed November 11,2017.

OECD. (2017). *United Kingdom, Trade and Investment Statistical Note.* http://www.oecd.org/investment/UNITED-KINGDOM-trade-investment-statistical-country-note.pdf. Accessed November 10, 2017.

Rothman, A. (2016). *Brexit Impact on China.* Advisor Perspectives.

Summers, T. (2016). *Brexit and the UK's China Challenge.* https://www.chathamhouse.org/expert/comment/.

Summers, T. (2017). *Brexit: Implications for EU-China Relations.* Chatham House: Royal Institute of International Affairs.

UK Trade & Investment. (2015). *UKTI Inward Investment Report 2014 to 2015* (online viewing). https://www.gov.uk/government/publications/ukti-inward-investment-report-2014-to-2015/ukti-inward-investment-report-2014-to-2015-online-viewing. Accessed November 2, 2017.

UK Trade and Investment Inward Investment Reports. (2017). *Department for International Trade, Inward Investment Results.* https://www.gov.uk/government/collections/dit-inward-investment-results. Accessed November 9, 2017.

UNCTAD/WIR. (2017). World Investment Report 2017. *Investment and the Digital Economy.* https://unctad.org/en/PublicationsLibrary/wir2017_en.pdf. Accessed November 11, 2017.

Voss, H. (2017). China's Outward FDI to Europe After Brexit. *China Policy Institute: Analysis.* https://cpianalysis.org/2017/05/26/chinas-outward-fdi-to-europe-after-brexit/. Accessed October 16, 2017.

Winders, S. (2016). *Would a Post-Brexit UK Be Better Able to Sign Free Trade Agreements with the Rest of the World?* The Bruges Group.

13

The Potential Impacts of Brexit on the Japanese Economy

Anna Maria Dzienis

13.1 Introduction

Japan is the world's third-largest economy and the EU's second-biggest trade partner in Asia. Since 2013, Japan and the EU have been negotiating an Economic Partnership Agreement (EPA), the conclusion of which is expected to bring further trade facilitation. However, the results of the 2016 UK referendum brought a new perspective to the debate. Moreover, many UK-based Japanese firms have started considering how to respond to new regulations and how to deal with risk connected to exchange rate fluctuations after the UK leaves the EU.

This chapter provides an analysis of the internal and external situation of Japan against the background of recent developments in the EU related to Brexit. The first section examines current economic policy in

A. M. Dzienis (✉)
Collegium of World Economy,
SGH Warsaw School of Economics, Warsaw, Poland
e-mail: anna.dzienis@sgh.waw.pl

A. M. Kowalski (ed.), *Brexit and the Consequences for International Competitiveness*, https://doi.org/10.1007/978-3-030-03245-6_13

Japan and sheds light on Japan's perception of "the UK issue". The second section serves two purposes: it describes Japan's foreign trade and reports on Japan's outward foreign direct investment (OFDI). The third section follows up on the previous one in the context of international agreements, with a focus on the Japan-EU Economic Partnership Agreement.

As the UK's House of Commons admits, much about the UK exit negotiations remains unclear and unknown: its mechanics, legal effects, policy consequences. *The biggest 'unknown' is what the withdrawal agreement and any other Brexit-linked agreements are going to contain.*[1]

13.2 Is Brexit Uncertainty Hitting Japan's Economic Policy?

The burst of the real estate bubble in the late 1980s derailed Japan from its path of dynamic economic growth. The worsening performance of Japan's economy exacerbated deflationary trends that have been present in the country since 1998. The Bank of Japan's governor, Haruhiko Kuroda, in a speech at the Council on Foreign Relations in New York on 10 October 2013 cited the example of the deflation situation in Japan and said that the minimum fare of JPY 160 for a subway ticket in Tokyo has not changed since 1995; while in New York, the charge has increased by over 60%.[2]

Subsequent government interventions were aimed at ending stagnation and stimulating the economy. Economic policy, however, lacked continuity, since Cabinets changed often. Shinzo Abe, who won the December 2012 general elections to sweep back into office as Japan's prime minister, has become widely recognised for his comprehensive economic policy package known as Abenomics and aimed at sustainably reviving the Japanese economy.

[1]Brexit Unknowns, Miller V. et al., 9 November 2016, House of Commons Library, p. 4, researchbriefings.files.parliament.uk/documents/…/CBP-7761.pdf. Accessed 11 May 2017.

[2]Kuroda, H., Governor of the Bank of Japan, Speech at the Council on Foreign Relations in New York, 10 October 2013, https://www.boj.or.jp/en/announcements/press/koen_2013/ko131010a.htm/. Accessed 1 November 2017.

Table 13.1 Japan's three "arrows" of Abenomics

Aggressive monetary policy	Flexible fiscal policy	Growth strategy
Quantitative easing	Increased public spending (stimulus package)	Structural reforms
• Open-ended asset purchasing • Buying long-term government bonds	• Disaster prevention and reconstruction • Stimulating private investment • Social and regional expenditures	• Tax reform • Innovation • Labour market (women) • Agriculture • Healthcare sector • Energy • Demography • International cooperation (trade and investment)
Stimulate the economy: – Lower interest rates – Weaker yen	 – Higher wages – Increase in exports	
Expectations of inflation and economic growth		

Source Own elaboration

13.2.1 Abenomics

Abe's economic policy includes aggressive monetary policy measures, flexible fiscal policy and growth strategy, the three so-called arrows of Abenomics. The expansionary monetary policy based on quantitative easing was designed to stimulate the economy through lower interest rates and a weaker yen. This, in turn, was expected to result in a rise in inflationary expectations and lead to economic growth. Simultaneously, a fiscal stimulus package aimed to bring about an increase in public investment, one of the key drivers of Japan's economic growth, and a growth strategy focused on crucial structural reforms (see Table 13.1).

As the International Monetary Fund stresses, Abenomics' third arrow—structural reforms—constitutes an essential element of putting Japan on a path of growth.[3] In particular, long-awaited labour market reforms are expected to increase productivity, investment and wages.

Additionally, Abe's economic policy relies on international trade and investment. Japan is promoting free trade and taking part in negotiations on a Regional Comprehensive Economic Partnership (RCEP) and a Japan-China-South Korea Free Trade Agreement (FTA). The country has reached an agreement with the EU on their bilateral Economic Partnership Agreement (EPA), thanks to which the partners would create a new economic zone accounting for about 28% of the world's GDP and 37% of global trade.[4] Moreover, Japan is still hopeful for the return of the USA to Trans-Pacific Partnership (TPP) negotiations, which US President Donald Trump has decided to leave, to the disappointment of Abe. TPP has been central to the Abe administration's economic and foreign policy strategies, since, according to the government, the Partnership would increase Japan's GDP by 2.59% and employment by 1.25% (Solis and Urata 2018).

13.2.2 Macroeconomic Setting

According to data by Japan's Cabinet Office, the Japanese economy is in moderate recovery.[5] The country's 2016 GDP growth rate was 1% year on year and is expected to reach 1.3% yoy in 2017 (IMF 2017).

The quarter-on-quarter real GDP growth rate has been positive for six quarters now, with private consumption growing gradually. Throughout 2016, the growth was driven by net exports (see Fig. 13.1).

---

[3]Country Report No. 17/242, IMF, p. 3. https://www.imf.org/en/News/Articles/2017/07/31/NA073117-For-Japan-Economy-Now-Is-the-Time-to-Step-Up-Reforms. Accessed 15 October 2017.

[4]Ibidem., p. 81 (Statement by Masaaki Kaizuka, Executive Director for Japan; Yoshihito Saito, Alternate Executive Director; and Masahiko Takeuchi, Advisor to Executive Director, 26 July 2017, p. 4).

[5]Cabinet Office, http://www5.cao.go.jp/keizai3/getsurei-e/2017oct.html. Accessed 15 October 2017.

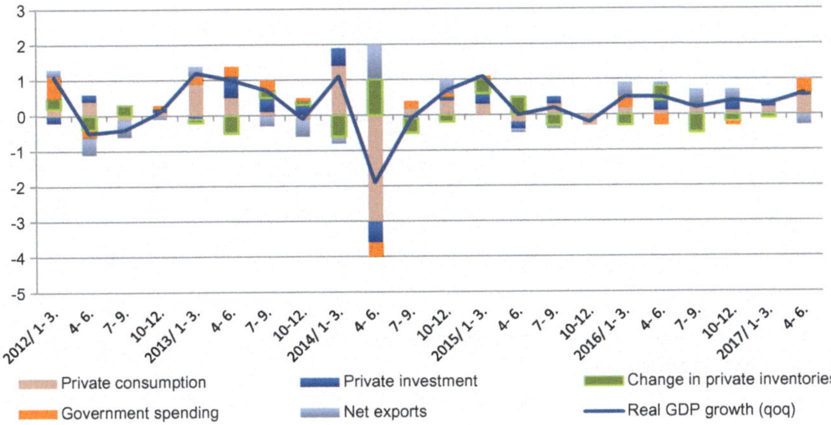

Fig. 13.1 Contributions to quarter-on-quarter GDP growth, seasonally adjusted (%, 2012–2017) (*Source* Cabinet Office, http://www.esri.cao.go.jp/en/sna/data/sokuhou/files/2017/qe172_2/gdemenuea.html. Accessed 15 October 2017. Own elaboration)

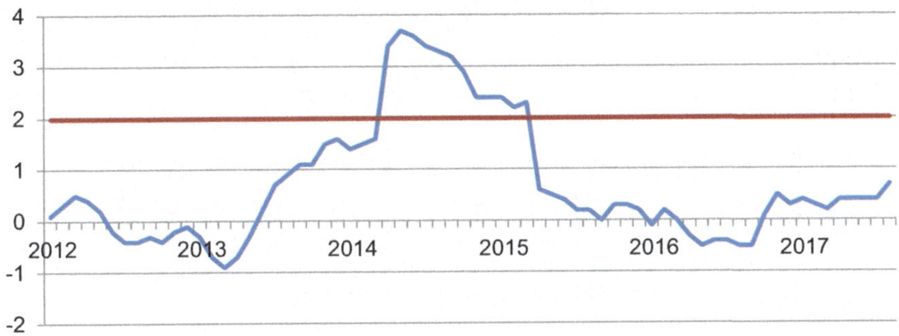

Fig. 13.2 Consumer price index, 2012–2017 (*Source* Statistics Bureau, www.stat.go.jp. Accessed 15 October 2017)

However, inflation remains low, far below the Bank of Japan's target of 2%. The only considerable rise in the consumer price index (CPI) in 2014 was triggered by a consumption tax hike, from 5% to 8% (see Fig. 13.2).

The labour market too has shown little sign of improvement. Although the unemployment rate fell to a 23-year low of 2.8% in

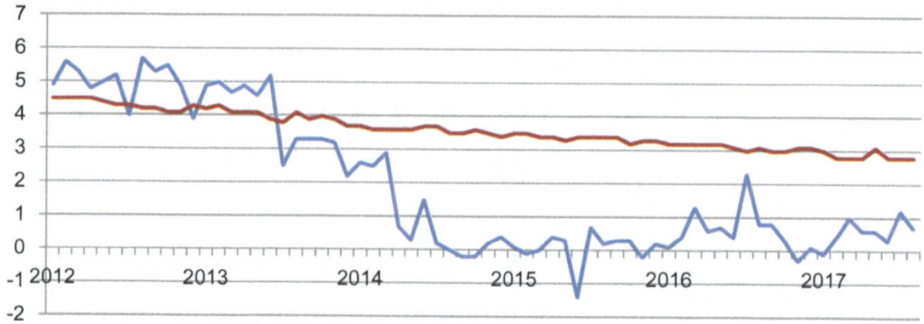

Fig. 13.3 Real wage index and unemployment rate, 2012–2017 (*Source* Statistics Bureau. Own elaboration)

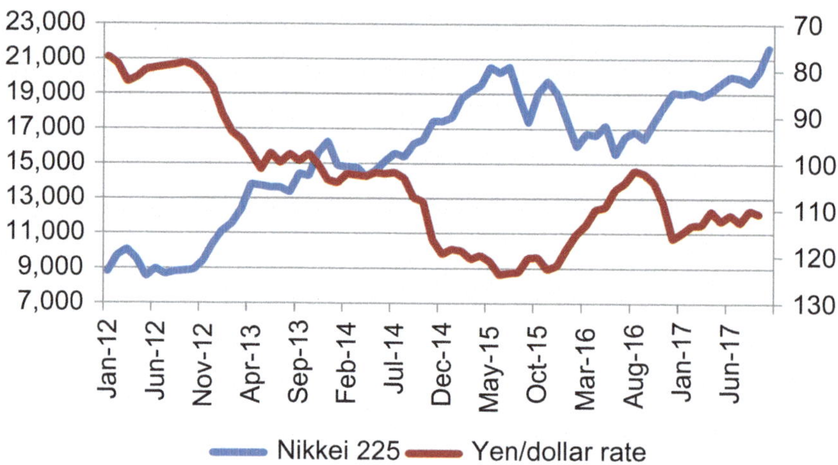

Fig. 13.4 Japanese yen against US dollar and the Nikkei index, 2012–2017 (*Source* Nikkei, https://indexes.nikkei.co.jp/nkave/index/profile?cid=7&idx=nk225. Accessed 20 October 2017. Own elaboration)

September 2017, the real wage index did not move accordingly (see Fig. 13.3).

Economic policies under Abenomics succeeded in bringing the yen to depreciate—the Japanese currency fell 50% against the dollar since the end of 2012. The depreciation lasted until mid-2016, when geopolitical instability and an appetite for the safe-haven yen among investors thwarted the government's efforts (see Fig. 13.4).

The authors of *The Impact of Brexit on the Japanese Economy*, a briefing paper on policy issues published by the National Diet Library, state that, under the risks of a fluctuating exchange rate, Japan is concerned that the yen might appreciate (*endaka*). They say that the *endaka* has a ripple effect on the economy: it diminishes company profits, decreases the competitiveness of exports, cuts capital investments and, in consequence, restrains employment. They also argue that such an appreciation could deal a devastating blow to Abenomics, which aims to ensure sound economic growth.[6]

13.2.3 Japan's Stance in the Brexit Discussion

Tokyo's reaction to the UK's decision to leave the EU was most clearly expressed in *Japan's Message to the United Kingdom and the European Union*, released by the country's Foreign Ministry in September 2016.[7] In the report, Japan voiced its concerns about the post-Brexit free trade system and overall economic uncertainty and stressed the need for a transition period. However, what the government in Tokyo seems most troubled about is the future of Japanese business in the UK. Japan explicitly stated:

> In light of the fact that a number of Japanese businesses, invited by the Government in some cases, have invested actively [in] the UK, which was seen to be a gateway to Europe, and have established value-chains across Europe, we strongly request that the UK will consider this fact seriously and respond in a responsible manner to minimize any harmful effects on these businesses. (Japan's Message 2016, p. 3)

[6] *The Impact of Brexit on the Japanese Economy* (英国のEU離脱と日本への経済的影響), Issue Brief No. 916, p. 10, August 2, 2016, http://dl.ndl.go.jp/view/download/digidepo_10159408_po_0916.pdf?contentNo=1. Accessed 15 October 2017.

[7] http://www.mofa.go.jp/files/000185466.pdf. Accessed 15 October 2017.

Table 13.2 Impacts of Brexit on Japan (short-term)

Impact on exports, imports and investments (if the economic situation in the UK and the EU aggravates)	Impact through exchange rate (if yen appreciates)
• Slump in exports due to weaker demand in the UK and the EU	• Decline in companies' profits: Value of exports, dividends and royalties in companies based abroad fall
• Decline in FDI in both countries as a result of growing uncertainty	• Drop in exports competitiveness

Source Issue Brief (2016, p. 10). Own elaboration

13.3 Impacts of Brexit on Japan's Trade and Investment

The possible implications of Brexit for Japan's trade and investment are listed in the briefing paper published by the National Diet Library (see Table 13.2).

Japan, as an export-reliant economy, could suffer from any new trade barriers. The government's efforts to create favourable conditions for future trade and investment between Japan and the EU, as well as Japan and the UK, were defined in the *Japan-UK Joint Vision Statement*, published in late August 2017, during British Prime Minister Theresa May's official visit to Japan. In the document, Japan confirmed that, in the event of the UK's exit from the EU, the Cabinet in Tokyo would work quickly to establish an economic partnership with the UK.[8]

13.3.1 Trade

Japan and the EU combined were responsible for around 40% of global trade in 2016. Japan had a significant surplus in trade with the EU until 2012, followed by a slight deficit in subsequent years. Meanwhile, bilateral trade between the UK and Japan has been growing since 2012, with

[8]*Japan-UK Joint Vision Statement*, August 2017, http://www.mofa.go.jp/mofaj/files/000285436.pdf. Accessed 3 November 2017.

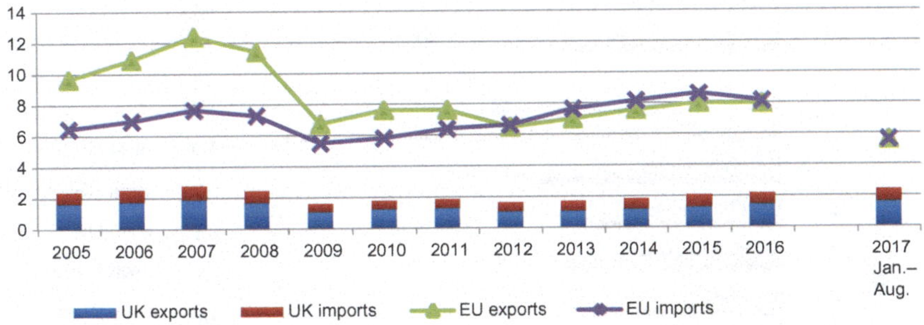

Fig. 13.5 Trade flows between the EU, UK and Japan, 2005–2017 (JPY trillion) (*Source* Ministry of Finance, Japan, http://www.mof.go.jp/. Accessed 3 November 2017. Own elaboration)

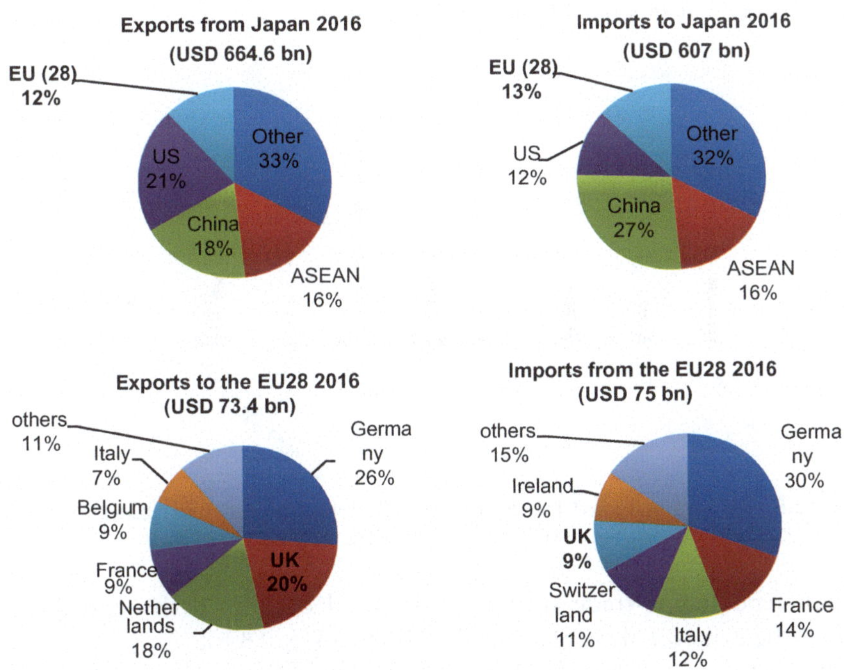

Fig. 13.6 The structure of Japan's foreign trade in 2016 (*Source* JETRO. Own elaboration)

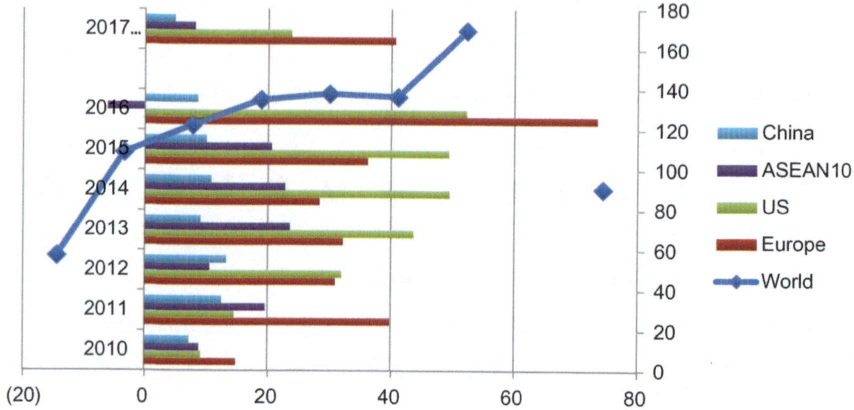

Fig. 13.7 Japan's outward FDI by region (balance of payment basis, net and flow; USD billion) (*Source* JETRO. Own elaboration)

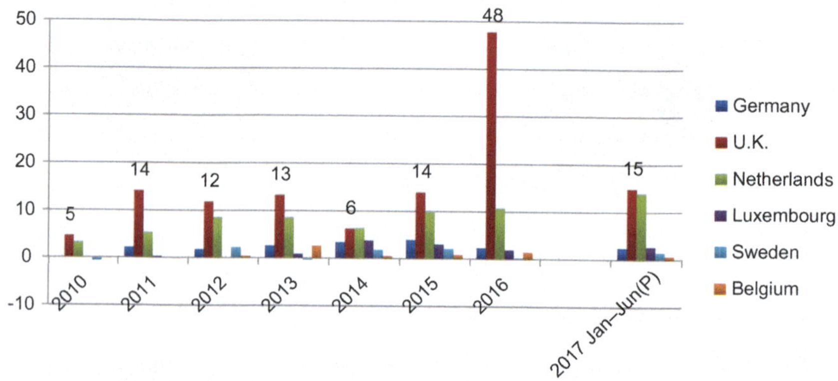

Fig. 13.8 Japan's outward FDI by country (balance of payment basis, net and flow; USD billion) (*Source* JETRO. Own elaboration)

Japan reporting a trade surplus. This trend continued through 2016, when the exit vote took place in the UK (see Fig. 13.5).

According to data published by the Japan External Trade Organisation (JETRO), the UK remained the second-largest export partner (USD 13.64 billion) and the fifth-largest import partner (USD 6.5 billion) for Japan in 2016 (see Fig. 13.6).

13.3.2 FDI

In August 2017, the London office of the Japan External Trade Organisation (JETRO) reported that FDI from Japan to the UK in 2016 was the highest ever (see Figs. 13.7 and 13.8).[9] On the one hand, many companies refrained from investing due to economic and political uncertainty surrounding the UK leaving the EU. On the other hand, there were numerous cases of companies strategically acquiring British firms. Moreover, after the British pound depreciated in the second half of 2016, there were many examples of companies seeking to gain from the exchange rate.

In terms of new FDI in the UK, 116 Japanese projects were reported, ranking Japan in fifth position after the USA, France, India and China.[10]

At the end of 2016, the UK accounted for 9% of Japan's total FDI stock worldwide, according to JETRO.

13.3.3 Japanese Companies in the EU

According to the Japanese Ministry of Foreign Affairs, there were 912 companies registered in the UK at the end of 2016, making the country the second-most popular destination for Japanese firms in Europe, after Germany.[11]

There are tight ties between UK-based Japanese companies and businesses operating in other EU countries. For example, 80–90% of the output of three big Japanese automotive producers active in the UK is exported to other EU members. In particular, a Nissan plant in Sunderland, north-eastern England, which is the largest car manufacturing facility in the UK, produces around 500,000 cars annually, or 10% of the company's total output (Issue Brief 2016, p. 11).

---

[9]JETRO, https://www.jetro.go.jp/biznews/2017/08/c319fe23c7f8d068.html. Accessed 3 November 2017.

[10]http://www.mofa.go.jp/mofaj/area/uk/data.html. Accessed 3 November 2017.

[11]http://www.mofa.go.jp/mofaj/area/uk/data.html. Accessed 3 November 2017.

In August 2017, the *Asian Nikkei Review* reported that Nissan Motor would increase its production in Sunderland by 20%, to around 600,000 cars a year, *aiming to offset higher costs expected after Britain's exit from the European Union becomes final.* The *Review* notes that British leader Theresa May has assured Nissan of London's assistance in maintaining the Sunderland factory competitive. Toyota Motor has also announced plans to invest at least USD 309 million to upgrade its British plant.[12]

Meanwhile, Hitachi Newton Aycliffe, Hitachi's rail vehicle manufacturing facility, whose production significantly depends on exports and imports, is reportedly concerned over the consequences of Brexit regarding free access to the European market (Issue Brief 2016, p. 11). The Bloomberg news agency quoted the company's chairman, Hiroaki Nakanishi, as saying during the World Economic Forum in Davos, Switzerland, that Hitachi may have some troubles when it becomes more reliant on exports after 2019.[13]

On 5 December 2016, JETRO's Europe, Russia and CIS Department published the results of the 2016 JETRO Survey on Business Conditions of Japanese Companies in Europe.[14] The organisation said that the survey had been conducted to understand the challenges that Japan-affiliated companies in Europe could face. The study was carried out in the third-quarter of 2016 among 1403 Japanese companies, with a response rate of 71% (JETRO 2016, p. 2).

When asked about 2016 operating profit forecasts, most companies in Europe reported an *increase* in profit. However, the response rate for profit growth among UK-based Japan-affiliated companies dropped by 0.7 p.p. year on year. Compared to 2016, 2017 operating profit forecasts showed the following responses: 31.6% of those polled said their profits would *increase*, 55.1% said they would *remain the same*, and

---

[12]https://asia.nikkei.com/Business/Companies/Nissan-bolstering-UK-production-as-split-with-EU-looms. Accessed 3 November 2017.

[13]https://www.bloomberg.com/news/articles/2017-01-20/hitachi-says-brexit-may-hurt-u-k-rail-plant-once-exports-needed. Accessed 3 November 2017.

[14]JETRO, https://www.jetro.go.jp/en/news/releases/2016/a329b6907c73f177.html. Accessed 15 October 2017.

Table 13.3 Possible repercussions of Brexit on Japanese business in the UK

• Increase in customs duty; appearance of non-tariff barriers=>higher costs and more complicated procedures
• Restrictions to labour migration=>problems with securing human resources
• Changes in the system of intellectual property rights, competition law, regulatory criterion and certification scheme=>necessity to deal with new regulations and legislation
• European Medicines Agency (EMA) relocation, fewer opportunities for research cooperation with other EU members, loss of EU funds for research=>loss of personnel, lower research and development capacity, erosion of industrial clusters
• Exchange rate fluctuations, economic stagnation, shrinking of the market, change in tax system, prolongation of political disarray

Source Issue Brief (2016, p. 12). Own elaboration

13.3% said they expected their profits to *decrease*. These responses moved the UK to second place from the bottom. The survey states clearly that the sharp decline in *increase* responses (by 9.1 p.p.) among UK-based companies stems from the UK's EU referendum results. At the same time, the number of *remain the same* responses rose by 3.8 p.p. (JETRO 2016, p. 4).

The three most frequent replies to the question about operational challenges in Europe were: (1) *European political and social conditions* (458 respondents, 47.9%), up from fourth position a year earlier; (2) *exchange rate fluctuations* (457 respondents, 47.8%), up from fifth a year earlier; (3) *securing human resources* (457 respondents, 47.8%); and (4) *high labour costs* (394 respondents, 41.2%). JETRO said that companies could feel uneasy about developments such as growing populism and EU scepticism as well as the impact of terrorism and a migration crisis on consumption and tourism. The organisation added that many Japan-affiliated companies underlined that *the uncertainty associated with the UK withdrawing from the EU* had already affected trade policy, investment policy, tax systems, labour issues and regulations (p. 14). This uncertainty seems to be justified by studies such as Golab et al. (2018) arguing that Germany and the UK influence all the markets. The studied firms said that, after the UK's exit from the EU, they would have to deal with various changes in regulations. Although the business outlook for the next year or two turned out not to have been affected so far, the results of the UK vote have started to influence the strategies of Japanese companies (JETRO 2016, p. 5).

To sum up, the following possible repercussions of Brexit on Japanese business in the UK can be named (see Table 13.3).

Apart from manufacturing and non-manufacturing companies, there are also big Japanese financial institutions in the UK. Two such institutions, Sumitomo Mitsui Financial Group Inc. and Nomura Holdings Inc., were in July 2017 reported to have taken the first steps to establish subsidiaries in Germany's Frankfurt to safeguard their operations in the EU after Brexit.[15] In *Japan's Message to the United Kingdom and the European Union*, the government in Tokyo asks that the UK and the EU maintain freedom of establishment and provision of financial services, including the *single passport* system (Japan's Message 2016, p. 4). Japanese analysts are convinced that, if the UK quits the EU, there will be a new agreement between the partners, but long negotiations and overall uncertainty will affect the Japanese economy.

13.4 Potential Consequences of Brexit for Japan's Political Relations

Abe's Cabinet, in its *National Security Strategy* of 17 December 2013, reaffirmed Europe's role in Japan's policy-making process: *Europe has the influence to formulate international public opinions, the capacity to develop norms in major international frameworks and a large economy. Japan and European countries, especially the United Kingdom, France, Germany, Italy, Spain and Poland, share universal values of freedom, democracy, respect for fundamental human rights and the rule of law, and principles such as market economy. They are partners for Japan, which together take a leading role in ensuring the peace, stability and prosperity of the international community … To effectively address global challenges, and to accomplish Japan's initiatives for a peaceful and prosperous international community, Japan*

[15] *The Japan Times*, https://www.japantimes.co.jp/news/2017/07/17/business/brexit-approaches-japanese-financial-firms-cementing-footholds-europe/. Accessed 5 November 2017.

will further strengthen its relations with Europe, including cooperation with the European Union (EU), the North Atlantic Treaty Organization (NATO), and the Organization for Security and Co-operation in Europe (OSCE).[16]

In this context, it seems crucial for Japan to ensure minimal disruption from Brexit and to fully cooperate so that the UK's exit from the EU would be negotiated smoothly.

13.4.1 EU and Japan: Major Agreements

According to the European Commission, the EU has concluded four major agreements with Japan[17]:

- EU-Japan Mutual Recognition Agreement (2001)
- Agreement on Cooperation on Anticompetitive Activities (2003)
- The Science and Technology Agreement (2009)
- The Agreement on Cooperation and Mutual Administrative Assistance (2008).

All these documents were signed to facilitate trade and overall cooperation between Japan and the UK. It is worth mentioning that the 2001 EU-Japan Mutual Recognition Agreement was the first bilateral agreement on mutual recognition for Japan.

13.4.2 Strategic Partnership Agreement (SPA)

The EU's Strategic Partnership with Japan, established in 2001, is—as the European Union External Action website defines it—"*a legally binding pact covering not only political dialogue and policy cooperation, but also cooperation on regional and global challenges, including environment*

[16] *National Security Strategy*, 17 December 2013, p. 26. https://www.cas.go.jp/jp/siryou/131217anzenhoshou/nss-e.pdf. Accessed 5 November 2017.

[17] http://ec.europa.eu/trade/policy/countries-and-regions/countries/japan/. Accessed 5 November 2017.

and climate change, development policy and disaster relief, and security policy".[18]

13.4.3 Economic Partnership Agreement (EPA)

The *National Security Strategy* also affirms that promoting economic partnership is essential for Japan's economic prosperity. The document mentions not only the EU-Japan Economic Partnership Agreement (JEEPA), but also the Trans-Pacific Partnership, the Free Trade Agreement among Japan, China and the ROK, and the Regional Comprehensive Economic Partnership (RCEP) (National Strategy 2013, p. 33).

The EPA has been negotiated since 2013, and its purpose is to lift trade barriers such as high Japanese customs taxes, e.g. nearly 40% on beef and up to 40% on cheese, and strict rules, such as long and costly procedures to approve each variety of fruit (European Commission).

The European Commission document *EU-Japan Economic Partnership Agreement* says that EU businesses expect the agreement to bring concrete benefits to European exporters:

- Removal of customs duties would save up to EUR 1 billion annually;
- EU exports of processed food to Japan could rise by up to 180%;
- EU exports of chemicals to Japan could rise by over 20%.

The document says that, since investment is crucial for the EU, member countries request that the Commission improve access to the Japanese market and upgrade rules to promote and protect EU investors in Japan.[19]

The Japan-EU Economic Partnership Agreement (JEEPA) negotiations—prompted by the unexpected withdrawal of Trump's

---

[18]European Union External Action, https://eeas.europa.eu/headquarters/headquarters-homepage/19223/eu-japan-political-relations_en. Accessed 5 November 2017.

[19]EU-Japan Economic Partnership Agreement, 1 July 2017, pp. 3, 5. http://trade.ec.europa.eu/doclib/docs/2017/july/tradoc_155684.pdf. Accessed 5 November 2017.

administration from the Trans-Pacific Partnership (TPP) talks—were concluded in July 2017 when the two parties, Japan and the EU's 28 member countries, reached an *agreement in principle* on their future economic relations. During a press conference, Japan's government spokesman Norio Maruyama said that *with regard to the Japan-EU EPA, Japan will request the UK's cooperation for its early signing and entry into force* (MOFA).[20] The agreement is expected to enter into force in 2019. It is hypothetical but, if the UK is still a member when the Japan-EU EPA is ratified, the country may be able to benefit from the agreement. The so-called *cars-for-cheese* trade deal provides for tariff reductions over the next 15 years, facilitation of EU food and agricultural product exports to Japan, and easier access to the European market for Japanese cars. However, the current provisions do not include any investment protection scheme, as the partners could not settle on an investment dispute resolution mechanism. The European Commission proposed its revised Investment Court System (ICS) and clarified that *there can be no return to the old-style investor-state dispute settlement system* (ISDS) (EC),[21] while Japan opted for the latter (Mizuho Insaito, p. 4).[22]

According to the JETRO 2016 survey, concluding the EPA/FTA could bring *major advantages* to Japan-affiliated companies. The firms believe that, thanks to these agreements, *improvement of price competitiveness* would be achieved (JETRO 2016, p. 6). Moreover, the survey data shows that the EU-Japan EPA would be of great importance to industries such as motor vehicle and motorcycle parts and accessories, general machinery (including metal moulds and machine tools), motor vehicles and motorcycles. However, Japan is thought to have immense trade potential for the EU as it is considered to be one of the least penetrated markets in the OECD (Bungenberg and Hazarika 2018).

---

[20]http://www.mofa.go.jp/press/kaiken/kaiken4e_000414.html. Accessed 15 January 2018.

[21]http://trade.ec.europa.eu/doclib/docs/2017/july/tradoc_155693.doc.pdf. Accessed 15 January 2018.

[22]日EU・ EPA交渉大枠合意の意義。日本の通商戦略の再起動、みずほインサイト、みずほ総合研究所、, https://www.mizuho-ri.co.jp/publication/research/pdf/insight/pl170712.pdf. Accessed 15 January 2018.

Japan's government has prioritised eliminating tariffs on industrial products (including 10% on motor vehicles, and up to 14% on electrical machinery) and improving regulatory issues encountered by Japanese companies in Europe on Non-Tariff Measures (NTMs). Japan has named three areas of special significance to the Japan-EU EPA:

- Reviving economic growth (increase in inward FDI; rise in exports and outward FDI; internationalisation of Japanese companies);
- Strengthening relations with the EU (job creation, boosting trade and FDI);
- Benefitting from the international position (free trade as an alternative to protectionist policies; competitiveness).[23]

13.5 Conclusions

The UK's decision to leave the European Union has triggered many reactions in Japan. On the one hand, economists have focused on Brexit's impact on the Japanese economy, trying to avoid worst-case scenarios. On the other hand, in 2016 Japanese firms made record investments in the EU and the UK. Many investors have reportedly profited from exchange rate fluctuations and invested in British companies. And although this situation creates new business opportunities, there are still too many questions concerning the details of the UK exit negotiation process.

The UK has become the second-largest safe harbour for Japanese businesses in Europe, after Germany. This gives Japan a strong position in relations with EU partners and Japan's voice in the Brexit discussion cannot be left unheard. Japan has already requested measures such as maintenance of the current tariff rates and customs clearance procedures, access to non-UK or non-EU workers, and the maintenance of the "single passport" system. It is possible to imagine a situation in which decisions by Japanese investors on the future of their firms will affect the British economy.

[23]http://www.mofa.go.jp/mofaj/files/000013819.pdf. Accessed 5 November 2017.

References

Bloomberg. https://www.bloomberg.com/news/articles/2017-01-20/hitachi-says-brexit-may-hurt-u-k-rail-plant-once-exports-needed.

Brexit Unknowns, Miller, V., et al. (2016, November 9). House of Commons Library. researchbriefings.files.parliament.uk/documents/.../CBP-7761.pdf.

Bungenberg M., & Hazarika A. (2018). The European Union's Trade and Investment Policy in Asia: New Challenges and Opportunities in a Changing Global Environment—Or: Following Individual Roadmaps. *Asia Europe Journal,* 15(4), 377–397.

Cabinet Office. http://www5.cao.go.jp/keizai3/getsurei-e/2017oct.html and http://www.esri.cao.go.jp/en/sna/data/sokuhou/files/2017/qe172_2/gdemenuea.html.

Council on Foreign Relations. https://www.cfr.org/backgrounder/abenomics-and-japanese-economy and https://www.cfr.org/event/overcoming-deflation-bank-japans-challenge.

European Commission. http://ec.europa.eu/trade/policy/countries-and-regions/countries/japan/, http://ec.europa.eu/trade/policy/in-focus/eu-japan-economic-partnership-agreement/ and http://trade.ec.europa.eu/doclib/docs/2017/july/tradoc_155693.doc.pdf.

European Union External Action. https://eeas.europa.eu/headquarters/headquarters-homepage/19223/eu-japan-political-relations_en.

Golab, A., Jie, F., Powell, R., & Zamojska, A. (2018). Cointegration Between the European Union and the Selected Global Markets Following Sovereign Debt Crisis. *Investment Management and Financial Innovations, 15*(1), 35–45.

International Monetary Fund. (2017, July). (Country Report No. 17/242). https://www.imf.org/en/News/Articles/2017/07/31/NA073117-For-Japan-Economy-Now-Is-the-Time-to-Step-Up-Reforms.

Japan's Message. (2016, September). Ministry of Foreign Affairs. www.mofa.go.jp/files/000185466.pdf.

Japan-UK Joint Vision Statement. (2017, August). http://www.mofa.go.jp/mofaj/files/000285436.pdf.

JETRO. (2016). JETRO Survey on Business Conditions of Japanese Companies in Europe, December 2016, Japan External Trade Organization (JETRO), Overseas Research Department, Europe, Russia and CIS Division. https://www.jetro.go.jp/ext_images/en/reports/survey/pdf/rp_firms_euro201612.pdf, https://www.jetro.go.jp/biznews/2017/08/c319fe23c7f8d068.html.

Kuroda, H. (2013, October 10). *Governor of the Bank of Japan, Speech at the Council on Foreign Relations in New York*. https://www.boj.or.jp/en/announcements/press/koen_2013/ko131010a.htm/.

Ministry of Economy, Trade and Industry. http://www.meti.go.jp/main/60sec/2016/20160628001.html.

Ministry of Finance, Japan. http://www.mof.go.jp/.

Ministry of Foreign Affairs, Japan. http://www.mofa.go.jp/mofaj/area/uk/data.html, http://www.mofa.go.jp/files/000185466.pdf, http://www.mofa.go.jp/mofaj/files/000013819.pdf, and http://www.mofa.go.jp/press/kaiken/kaiken4e_000414.html.

Mizuho Research Institute. 日EU· EPA交渉大枠合意の意義。日本の通商戦略の再起動、みずほインサイト、みずほ総合研究所、. https://www.mizuho-ri.co.jp/publication/research/pdf/insight/pl170712.pdf.

National Security Strategy. (2013, December 17). https://www.cas.go.jp/jp/siryou/131217anzenhoshou/nss-e.pdf.

Nikkei. https://indexes.nikkei.co.jp/nkave/index/profile?cid=7&idx=nk225.

Nikkei Asian Review. https://asia.nikkei.com/Business/Companies/Nissan-bolstering-UK-production-as-split-with-EU-looms.

Nikkei Shimbun. https://www.nikkei.com/article/DGXLASGM09H5Y_Z00C17A2FF1000/.

Solis, M., & Urata, S. (2018). Abenomics and Japan's Trade Policy. *Asian Economic Policy Review*, *13*, 106–123.

Statistics Bureau. www.stat.go.jp.

The Government of Japan. https://www.japan.go.jp/abenomics/index.html.

The Impact of Brexit on the Japanese Economy. (2016, August 2). [英国のEU離脱と日本への経済的影響] (Issue Brief No. 916, p. 10). http://dl.ndl.go.jp/view/download/digidepo_10159408_po_0916.pdf?contentNo=1.

The Japan Times. https://www.japantimes.co.jp/news/2017/07/17/business/brexit-approaches-japanese-financial-firms-cementing-footholds-europe/.

14

Brexit and Sterling Depreciation: Impact on Selected Economies

Anna Sznajderska

14.1 Introduction

The already observed short-term impacts of Brexit include a depreciation of the pound sterling, stock market declines and a freeze in investment. Meanwhile, various estimates of the long-term effect of Brexit on the UK's GDP point to a significant reduction in the UK's GDP growth rate (see Fig. 14.1 and Chang 2017, p. 11). One of the main concerns is the need to negotiate new trade deals and possible barriers in trade with the EU, which would not necessarily be replaced by greater free trade elsewhere. Another important aspect is the uncertainty that the Brexit vote has created in markets worldwide. Some companies could be delaying investment decisions, while awaiting details on what a Brexit deal would mean for tariffs and other trade barriers (Andy Palmer, Tokyo, August 2017).

A. Sznajderska (✉)
Collegium of World Economy,
SGH Warsaw School of Economics, Warsaw, Poland
e-mail: asznajd@sgh.waw.pl

A. M. Kowalski (ed.), *Brexit and the Consequences for International Competitiveness*, https://doi.org/10.1007/978-3-030-03245-6_14

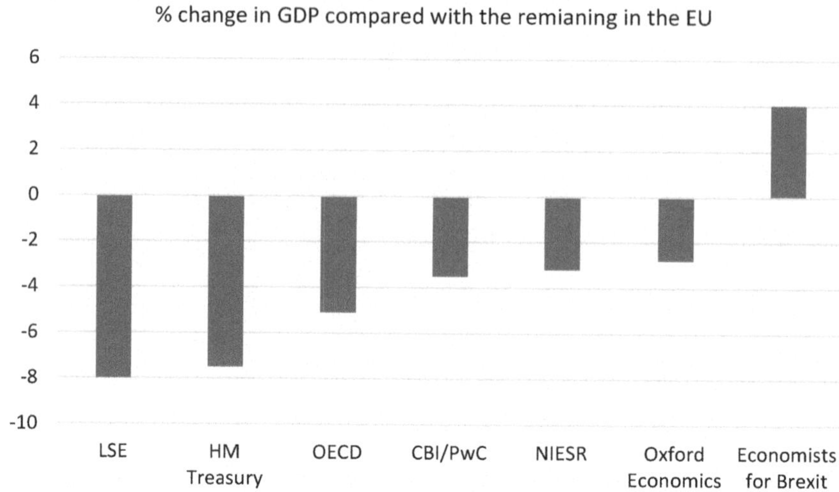

Fig. 14.1 Estimates of long-term effect of Brexit on national income (*Source* own compilation based on Giles [2016])

In this study, however, we concentrate on one of the most dramatic and already visible economic consequences of the Brexit decision. On the night of the decision itself, the pound suffered its biggest one-day fall against the dollar. On 27 June 2016, the pound fell to trade at 1.315 against the dollar, hitting a 31-year low. The depreciation was a signal that investors' expectations about the UK's economic performance had deteriorated. It seems that investors were expecting that leaving the EU would impose a long-term and permanent economic cost on Britain.

In this study, we apply an advanced econometric tool known as the global vector autoregressive model (GVAR).[1] Economists from the *European Bank for Reconstruction and Development* (see EBRD 2016), for example, use the GVAR model to assess the impact of Brexit on economic growth in the EBRD region. They distinguish two scenarios, a soft Brexit, where trade relationships are kept at

[1]To the best of our knowledge, the GVAR methodology was not used to study the effects of the pound's depreciation after the Brexit vote in any other study.

close to the present level, and a hard Brexit, where trade relationships between the UK and the EU are significantly disrupted. Their results show that both scenarios decrease the level of economic growth in the EBRD region, where, under a hard Brexit, the impact is much more severe.

The aim of this chapter is to assess the impact of an unexpected depreciation of the sterling on economic growth and on stock price indices in the UK and in selected other economies. The main hypothesis is that a depreciating sterling has significant effects on both economic growth and stock price indices in these countries.

The structure of the chapter is as follows. Section 14.2 concerns the transmission of negative exchange shocks in the UK. Section 14.3 describes the global vector autoregressive model. Sections 14.4 and 14.5 present the results obtained from the GVAR model. Section 14.4 discusses the impact of the sterling's depreciation on GDP in the UK and selected other economies. Section 14.5 focuses on the impact of sterling depreciation on stock prices in the UK and in the selected economies. The last section concludes.

14.2 Transmission of Exchange Rate Shocks in the UK

The depreciation of the domestic currency makes imports more expensive while making exports cheaper. It should cause imports to fall while causing exports to rise. This, in turn, should cause domestic employment as well as wages to rise. But for that to happen, demand for exports must increase.

Theoretically, the impact of the exchange rate on the trade balance and, consequently, on aggregate demand, depends on the Marshall-Lerner condition. It states that currency depreciation will have a positive impact on the trade balance when the sum of the absolute values of price elasticity of exports and imports is greater than one. This is because one can distinguish between a quantity effect and a cost effect.

The quantity effect means that, after currency depreciation, customers buy more domestically produced goods. Both domestic customers buy fewer imported goods and foreign customers buy more exported goods, which generates a positive effect on the trade balance. The cost effect means that, because of higher import costs after currency depreciation, more may be spent on imports initially when consumption patterns remain the same. This generates a negative effect on the trade balance. If the quantity effect is greater than the cost effect, then the Marshall-Lerner condition is fulfilled, meaning an improvement in the balance of trade.

However, Aiello et al. (2015), for example, show that the long-run level of exports appears to be unrelated to the real exchange rate for the UK. As a consequence, a further depreciation of the sterling could lead to sharp price increases because of higher import prices with no offsetting effect for exports. Moreover, British exporters are highly integrated with global supply chains. According to OECD data, the import content of UK exports is around 23%, compared with around 15% for the USA and Japanese exports (see Skidelsky 2016). This means that British exporters need imported inputs. Because of an increase in import prices, also export prices are less competitive. As a result, economic growth in the UK is expected to decrease rather than increase if the sterling depreciates.

It is interesting to note that Paul Krugman argues that a weaker pound should not be viewed as an additional cost of Brexit but as a part of an adjustment. He writes that Britain is experiencing a version of what is known as Dutch disease. The London City's financial experts are crowding out manufacturing by keeping the currency strong. Thus, a weak pound helps British manufacturing. The UK faces the prospect of largely increased transaction costs between Britain and the rest of Europe, which creates an incentive to move financial services away from the smaller economy (the City of London) into the larger (Europe). Such a move can be prevented by paying lower wages and therefore increasing competitiveness. In effect, the UK needs a weaker currency to offset the adverse impact of its smaller market.

14.3 Research Methodology

We apply a global vector autoregressive (GVAR) model to assess the impact of depreciation of the pound on selected economies. The GVAR model comprises a compact model of the whole world economy designed to explicitly model economic and financial interdependencies at the national and international levels.

Originally, the global vector autoregressive model was proposed by Pesaran et al. (2004), and it was further developed by Dees et al. (2007). It is possible to use the GVAR model to investigate a number of different problems (see, e.g., Cesa-Bianchi et al. [2012], note 2).

We use the modified GVAR Toolbox 2.0, which contains the necessary procedures in Matlab and a user-friendly interface in Excel (see Smith and Galesi 2014).

The GVAR model consists of individual country vector error-correcting models that include both domestic and foreign variables. The foreign variables are constructed on the basis of trade and financial linkages between countries. The individual country models are linked together, and the model is solved for the world as a whole.

Therefore, estimation of the GVAR model is a two-step procedure. First, we estimate small VARX models for each country that are conditional on the rest of world. The country-specific models comprise domestic, foreign and optionally global variables or dominant unit variables. Second, by using the spillover matrix we link individual countries' models into one global VAR model.

Let us consider N countries. We define the following VARX * (P, R) model for country i:

$$x_{it} = \alpha_{i0} + \alpha_{it}t + \sum_{p=1}^{Pi} \Phi_{ip}x_{i,t-p} + \sum_{r=0}^{Ri} \Lambda_{ir}x^*_{i,t-r} + u_{it}, \quad (14.1)$$

where $x_{i,t}$ is vector $1 \times k_i$ of domestic variables, $x^*_{i,t}$ is vector $1 \times k^*_i$ of foreign variables, $x^*_{it} = \sum_{j=0}^{N} \omega_{ij}x_{jt}$, $\omega_{ii} = 0$, ω_{ij} are weights that are calculated on the basis of bilateral trade or financial flows matrix, $\sum_{j=0}^{N} \omega_{ij} = 1$.

The model can be written in the following error correction form:

$$\Delta x_{it} = \mu_i + \sum_{j=1}^{r_i} \gamma_{ij} ECT_{ij,t-1} + \sum_{p=1}^{P} \widetilde{\Phi}_{ip} \Delta x_{i,t-p} + \sum_{r=0}^{R} \widetilde{\Lambda}_{ir} \Delta x^*_{i,t-r} + e_{it},$$

where r_i is the number of cointegrating relations.

We define vector $z_{it} = \begin{pmatrix} x_{it} \\ x^*_{it} \end{pmatrix}$, which, for a given country, contains its domestic as well as foreign variables. We can rewrite the model as:

$$A_{i0} z_{it} = a_{i0} + a_{i1} t + A_{i1} z_{i,t-1} + \cdots + A_{ip} z_{i,t-Pi} + u_{it},$$

where $A_{i0} = \left(I_{k_i}, -\Lambda_{i0}\right)$, $A_{ij} = \left(\Phi_{ij}, \Lambda_{ij}\right) j = 1, \ldots \max(P_i, R_i)$, $\Phi_{ij} = 0$ for $j > P_i$ and $\Lambda_{ij} = 0$ for $j > R_i$, $z_{it} = W_i x_t$, where W_i are $\left(k_i + k^*_i\right) \times k \left(k = \sum_{i=0}^{N} k_i\right)$ link matrices calculated on the basis of trade flows and $x_t = \left(x'_{0t}, x'_{1t}, \ldots, x'_{Nt}\right)'$. Further, the model can be written as:

$$A_{i0} W_i x_t = a_{i0} + a_{i1} t + A_{i1} W_i x_{t-1} + \cdots + A_{ipi} W_i x_{t-pi} + u_{it},$$

Finally, by stacking the individual country models, we arrive at the global VAR model with domestic variables only:

$$G_0 x_t = a_0 + a_1 t + G_1 x_{t-1} + \cdots + G_p x_{t-p} + u_t, \qquad (14.2)$$

$$G_0 = \begin{pmatrix} A_{00} W_0 \\ A_{10} W_1 \\ \cdots \\ A_{N0} W_N \end{pmatrix}, G_j = \begin{pmatrix} A_{0j} W_0 \\ A_{1j} W_1 \\ \cdots \\ A_{Nj} W_N \end{pmatrix}, a_0 = \begin{pmatrix} a_{00} \\ a_{10} \\ \cdots \\ a_{N0} \end{pmatrix},$$

$$a_1 = \begin{pmatrix} a_{01} \\ a_{11} \\ \cdots \\ a_{N1} \end{pmatrix}, u_t = \begin{pmatrix} u_{0t} \\ u_{1t} \\ \cdots \\ u_{Nt} \end{pmatrix}.$$

G_0 is known from the estimation of individual country models. We thus multiply both sides of Eq. (14.2) by G_0^{-1} and we get the GVAR(P) model:

$$x_t = b_0 + b_1 t + F_1 x_{t-1} + \cdots + F_p x_{t-P} + \varepsilon_t, \qquad (14.3)$$

where $b_0 = G_0^{-1} a_0$, $b_1 = G_0^{-1} a_1$, $F_j = G_0^{-1} G_j$ $j = 1, \ldots, p$, $\varepsilon_t = G_0^{-1} u_t$. Equation (14.3) is solved recursively.

After estimating the GVAR model, generalised impulse response functions (GIRFs) are calculated. It is important to note that, because of a large number of variables, it is difficult to use standard impulse response functions that assume orthogonal shocks (see Sims 1980). GIRFs were introduced by Koop et al. (1996). The shape of the GIRFs does not depend on the ordering of the variables. The GIRFs may be represented by the following equation:

$$\text{GIRF}(x_t, n, \varepsilon_{jlt}) = E\left[x_{t+n} | \varepsilon_{jlt} = \sqrt{\sigma_{jj,ll}}, I_{t-1}\right] - E[x_{t+n} | I_{t-1}] \qquad (14.4)$$

where I_{t-1} is an information set at time $t-1$, $\sigma_{jj,ll}$ is the diagonal element of the variance–covariance matrix Σ_ε corresponding to the lth equation in the jth country and n is the horizon.

Our sample consists of 55 economies (see Table 14.1). The economies together cover more than 90% of global GDP. When deciding on the choice of countries, in the first step we take all the countries included in the BIS effective exchange rate indices—60 economies plus the euro area (broad weights). We end up, however, with 55 economies, because we notice that including Algeria, Chinese Taipei, Malta, the United Arab Emirates and Venezuela makes the model unstable, which is probably due to low quality of data for these countries.

The euro-area countries are grouped into the euro-area region. We use quarterly observations. The data span is from 1995Q1 to 2016Q3. The main data used in the model are real GDP, the price level (CPI), the stock market index, the real effective exchange rate (REER) and the short-term interest rate for each country. We complement the data for domestic economies with the level of oil prices to take into account the situation in commodity markets. Economic ties between countries are approximated by bilateral flows of exports and imports of goods that are available on an annual basis. The matrices of trade flows are constructed

Table 14.1 Countries and regions included in the GVAR model

Euro area	Argentina	Malaysia
Austria	Australia	Mexico
Belgium	Brazil	New Zealand
Cyprus	Bulgaria	Norway
Estonia	Canada	Peru
Finland	Chile	Philippines
France	China	Poland
Germany	Colombia	Romania
Greece	Croatia	Russia
Ireland	Czech Republic	Saudi Arabia
Italy	Denmark	Singapore
Latvia	Hong Kong	South Africa
Lithuania	Hungary	Sweden
Luxembourg	Iceland	Switzerland
Netherlands	India	Thailand
Portugal	Indonesia	Turkey
Slovakia	Israel	United Kingdom
Slovenia	Japan	United States
Spain	South Korea	

on the basis of International Monetary Fund statistics, namely the *Direction of Trade Statistics (DOTS)*. The sources of the data used in the model are described in detail in Sznajderska (2018).

14.4 The Impact of Sterling Depreciation on GDP in the UK and Selected Other Economies

Below we present an impulse response analysis for the estimated GVAR model. The impulse response functions, which are for instance presented in Figs. 14.2, 14.4 and 14.5, refer to the time profile of the effects of sterling depreciation on all the variables in the model. The horizontal axis shows quarters after the shock.

We analyse a negative one-standard deviation shock to the real effective exchange rate (REER) in the UK (see Fig. 14.2), which corresponds to a 1.47% decrease in the REER at the time of impact.

The obtained results show that an unexpected depreciation of the sterling could have statistically and economically significant effects on

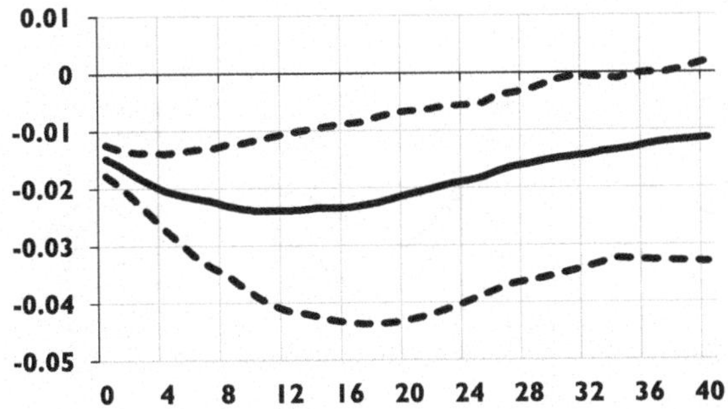

Fig. 14.2 The shock—sterling depreciation (*Note* Bootstrap mean estimates with 90% bootstrap error bounds)

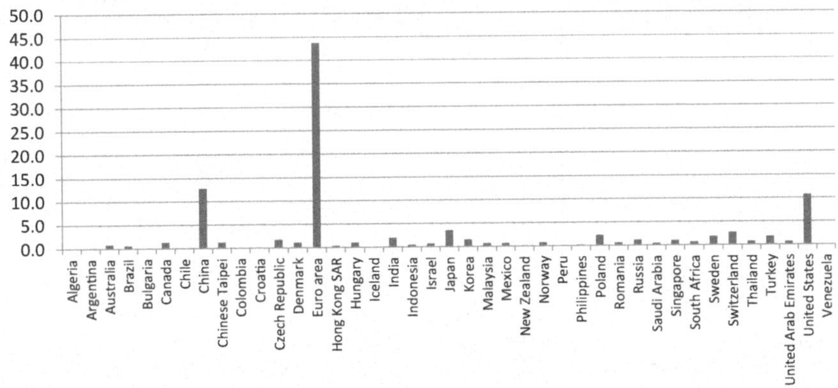

Fig. 14.3 BIS trade weights for UK (*Note* The trade weights are derived from manufacturing trade flows and capture both direct bilateral trade and third-market competition by double-weighting [see Klau and Fung 2006 for explanation of the weighting scheme])

other economies. Figure 14.4 shows the reaction of real GDP in the UK and in selected other economies.

After an unexpected depreciation of the real effective exchange rate in the UK, GDP declines in the majority of countries. In the UK, GDP decreases by a maximum 0.6% after 15 quarters. In other words, it can be easily calculated that a 1% depreciation of the REER would imply a 0.4% decrease in real UK GDP after 15 quarters. A negative effect of the depreciation on GDP is not obvious, because depreciation could have a positive effect on exports. But this effect seems to be dominated

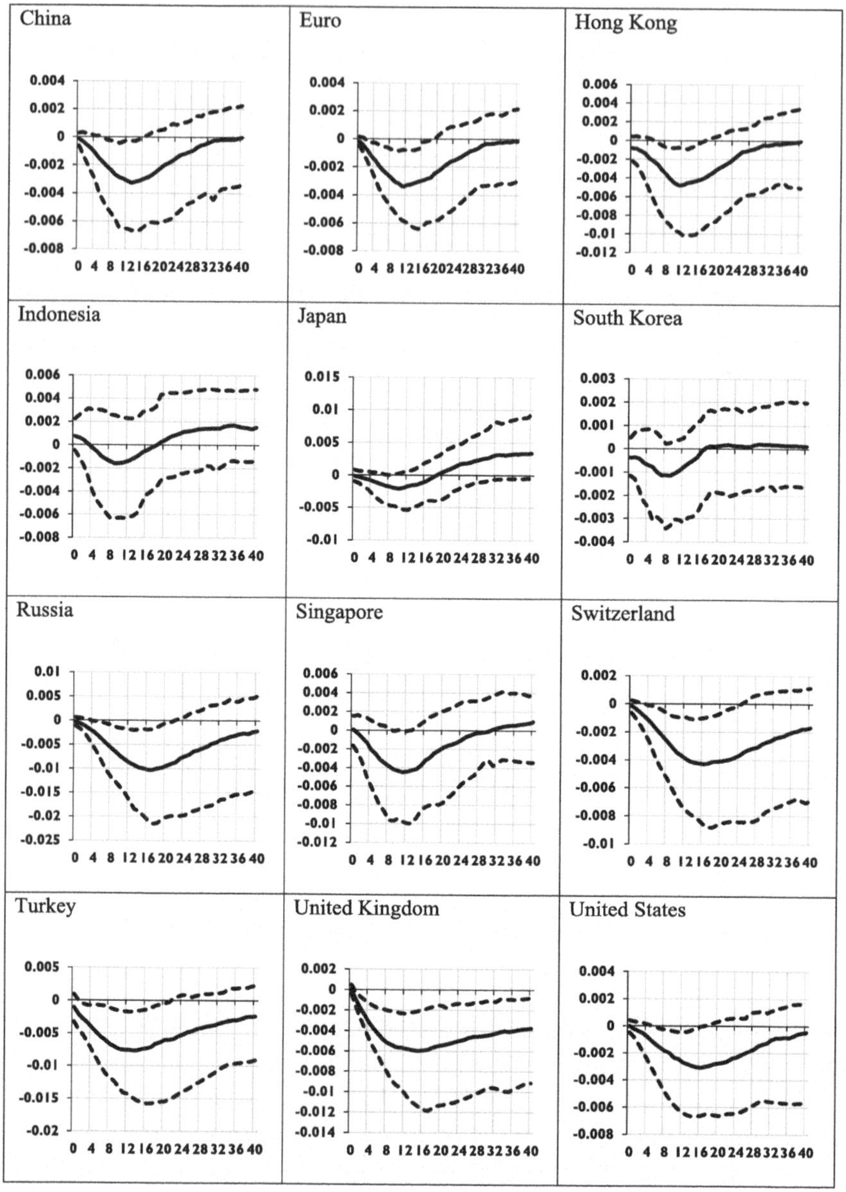

Fig. 14.4 Impulse responses of GDP to one-standard deviation shock to REER in UK (*Note* Bootstrap mean estimates with 90% bootstrap error bounds)

by more expensive imports of intermediates or increased interest rates, which weakens investment and consumption. Indeed, our results show a statistically insignificant increase in the level of prices and a statistically significant decrease in the level of interest rates in the UK.

In accordance with the obtained results, a depreciation of the REER in the UK (by 1.47% at the time of impact) causes a statistically significant reduction in real GDP in the euro area (by a maximum 0.3% after 12 quarters) and in the USA (maximum 0.3% after 16 quarters). The reaction of real GDP is statistically significant in countries such as China (maximum 0.3% after 13 quarters), Hong Kong (maximum 0.48% after 11 quarters), Russia (maximum 1% after 16 quarters), Singapore (maximum 0.4% after 11 quarters), Switzerland (maximum 0.4% after 16 quarters) and Turkey (maximum 0.76% after 14 quarters). On the other hand, the reaction of real GDP is statistically insignificant in countries including Indonesia, Japan and South Korea (see Fig. 14.4).

Figure 14.3 shows the trade links between the UK and the other economies. The euro area, China and the USA are the main trading partners of the UK. Our results indicate that sterling depreciation has statistically significant effects on these economies. The impact of an exchange rate shock on these economies seems to work through decreased domestic demand in the UK. It is worth noting that the effect is not significant for Japan and South Korea, both of which are among the UK's main trading partners (Fig. 14.4).

14.5 The Impact of Sterling Depreciation on Stock Prices in the UK and Selected Other Economies

In what follows, we consider the impact of the Brexit vote on stock prices. As a result of the Brexit decision, the FTSE 100 index was down 2.6% and the FTSE 250 (more closely tied to the UK economy) was down 7% on Friday, 24 June 2016. Many British companies were hit hard. But, as J. Treanor and K. Allen wrote in *The Guardian* on

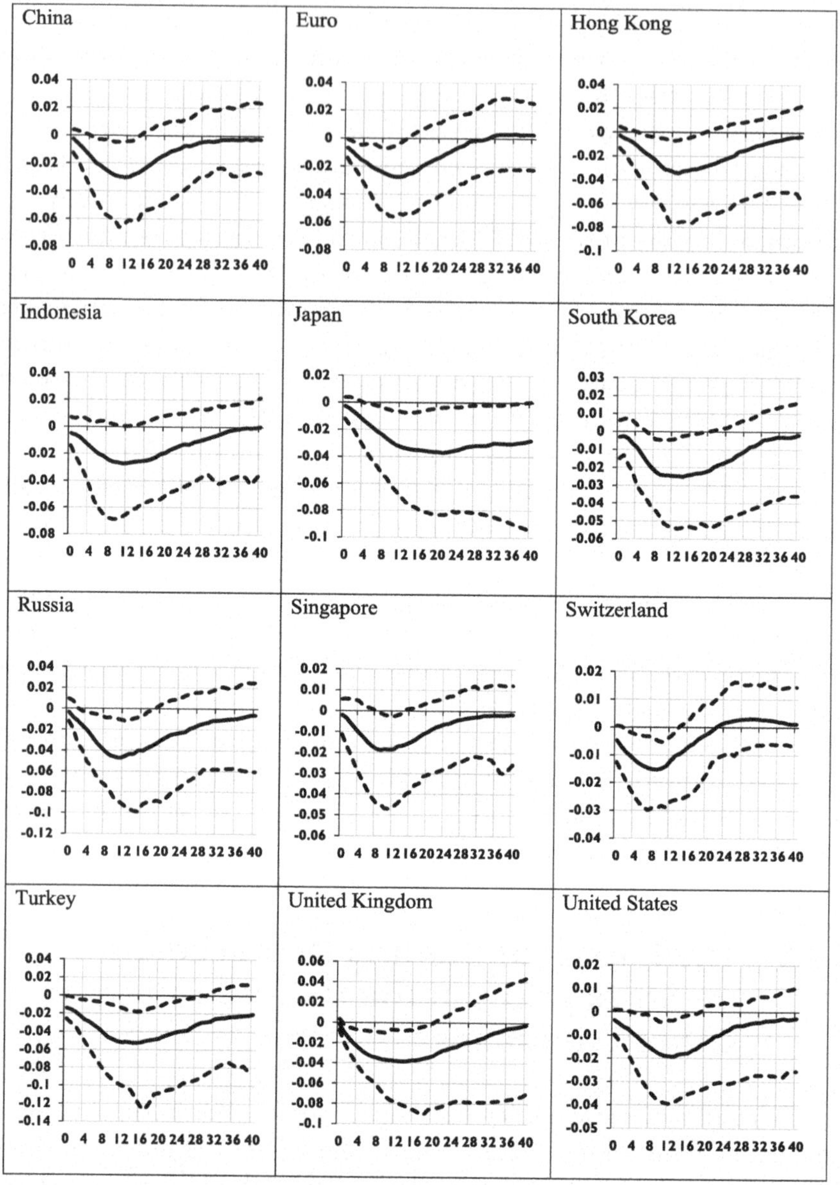

Fig. 14.5 Impulse responses of stock prices to one-standard deviation shock to REER in the UK (*Note* Bootstrap mean estimates with 90% bootstrap error bounds)

Monday, 27 June 2016: "*The fallout from the vote is being felt around the world. Italy's main index fell 4%, extending Friday's record losses of 12.5%. In Germany and France there were losses of 3%. At the time of the London close, on Wall Street the main share indices were all down more than 1%*". Thus, we study the reaction of stock prices to an unexpected exchange rate shock in the UK in selected other countries around the world.

Figure 14.5 shows the impulse response functions. First of all, a depreciating sterling causes a decrease in the stock price index in the UK. The stock price index in the UK decreases by a maximum 0.4% after 14 quarters following the analysed shock (see Fig. 14.2 for the shock).

As the result of the shock, stock price indices decrease in a statistically significant way in a number of countries, such as China (by a maximum 0.3% after 11 quarters), the euro area (by a maximum 0.27% after 11 quarters), Hong Kong (by a maximum 0.34% after 13 quarters), Japan (by a maximum 0.37% after 21 quarters), South Korea (by a maximum 0.25% after 14 quarters), Russia (by a maximum 0.47% after 11 quarters), Singapore (by a maximum 0.19% after nine quarters), Switzerland (by a maximum 0.15% after eight quarters), Turkey (by a maximum 0.5% after 15 quarters) and the USA (by a maximum 0.19% after 13 quarters). The reaction is not statistically significant for Indonesia. The results show high financial linkages between the UK stock market and stock markets in the other economies. This means that problems on the London Stock Exchange spill over greatly to many other financial markets, decreasing their competitiveness.

14.6 Conclusions

Brexit is "a major, significant financial shock" that could create "a whole bunch of economic, financial, political and also geopolitical uncertainties" (Nouriel Roubini, World Economic Forum in China). The spillover effects may appear all over the world. This is because the UK has strong trade linkages with other economies, on the one hand, and strong financial linkages, on the other, while the City of London is one of the world's largest financial centres.

In this chapter, we have discussed one of the consequences of Brexit, namely the depreciation of the British currency. We have analysed the effects of an unexpected exchange rate shock in the UK using the global vector autoregressive model.[2] The model enables concise analysis of the global economy as a whole. It takes into account economic linkages among a large number of economies.

Our results are in favour of the view that the UK leaving the EU will slow growth in Britain and reduce its competitiveness. Moreover, the results show that a further depreciation of the pound sterling will slow growth in a number of other countries, negatively affecting their competitiveness.

Also, an unexpected depreciation of the sterling appears to greatly affect financial markets all over the world. It could cause a decrease in the stock price index in the UK and most other economies included in our model.

References

Aiello, F., Bonanno, G., & Via, A. (2015). Again on Trade Elasticities: Evidence from a Selected Sample of Countries. *Eurasian Business Review, 5*(2), 259–287.

Cesa-Bianchi, A., Pesaran, M. H., Rebucci, A., & Xu, T. (2012). China's Emergence in the World Economy and Business Cycles in Latin America. *Economía, 12*(2, Spring), 1–75. Brookings Institution Press.

Chang, W. W. (2017). *Brexit and Its Economic Consequences.* Presented at the Brexit Roundtable Discussion, SUNY Buffalo, 10 April 2017.

Dees, S., Di Mauro, F., Pesaran, M. H., & Smith, L. V. (2007). Exploring the International Linkages of the Euro Area: A Global VAR Analysis. *Journal of Applied Econometrics, 22*(1), 1–38.

European Bank for Reconstruction and Development. (2016, November). *Regional Economic Prospects in EBRD Countries of Operations, Return to Modest Growth.*

[2]The weakness of this approach is the usage of generalised impulse response functions and not the standard impulse response functions that assume orthogonal shocks.

Giles, C. (2016, June 27). Brexit in Seven Charts—The Economic Impact. *Financial Times*.

Klau, M., & Fung, S. S. (2006). The New BIS Effective Exchange Rate Indices. *BIS Quarterly Review*.

Koop, G., Pesaran, M. H., & Potter, S. M. (1996). Impulse Response Analysis in Nonlinear Multivariate Models. *Journal of Econometrics, 74*(1), 119–147.

Pesaran, M. H., Schuermann, T., & Weiner, S. (2004). Modeling Regional Interdependencies Using a Global Error-Correcting Macroeconometric Model. *Journal of Business & Economic Statistics, 22,* 129–162.

Sims, C. (1980). Macroeconomics and Reality. *Econometrica, 48,* 1–48.

Skidelsky, R. (2016, October 21). Why Sterling's Collapse Is Not Good for the UK Economy. *The Guardian*.

Smith, L., & Galesi, A. (2014). *GVAR Toolbox 2.0* (1–181). University of Cambridge: Judge Business School. Available at https://sites.google.com/site/gvarmodelling/gvar-toolbox.

Sznajderska, A. (2018). The Role of China in the World Economy: Evidence from a Global VAR Model. *Applied Economics*. https://doi.org/10.1080/00036846.2018.1527464.

Index

A. M. Kowalski (ed.), *Brexit and the Consequences for International Competitiveness*, https://doi.org/10.1007/978-3-030-03245-6

The manufacturer's authorised representative in the EU is Springer Nature Customer Service Centre GmbH, Europaplatz 3, 69115 Heidelberg, Germany. If you have any concerns regarding our products, please contact ProductSafety@springernature.com

Printed and bound by CPI Group (UK) Ltd, Croydon, CR0 4YY
22/09/2026
02221063-0005